T0269027

Chaos

*A Mathematical Introduction*

AUSTRALIAN MATHEMATICAL SOCIETY LECTURE SERIES



1 Introduction to Linear and Convex Programming, N. CAMERON
2 Manifolds and Mechanics, A. JONES, A. GRAY & R. HUTTON
3 Introduction to the Analysis of Metric Spaces, J.R. GILES
4 An Introduction to Mathematical Physiology and Biology, J. MAZUMDAR
5 2-Knots and their Groups, J. HILLMAN
6 The Mathematics of Projectiles in Sport, N. DE MESTRE
7 The Peterson Graph, D. A. HOLTON & J. SHEEHAN
8 Low Rank Representations and Graphs for Sporadic Groups, C. PRAEGER & L. SOICHER
9 Algebraic Groups and Lie Groups, G. LEHRER (ed)
10 Modelling with Differential and Difference Equations, G. FULFORD, P. FORRESTER & A. JONES
11 Geometric Analysis and Lie Theory in Mathematics and Physics, A. L. CAREY & M. K. MURRAY (eds)
12 Foundations of Convex Geometry, W. A. COPPEL
13 Introduction to the Analysis of Normed Linear Spaces, J. R. GILES
14 The Integral: An Easy Approach after Kurzweil and Henstock, L. P. YEE & R. VYBORNY
15 Geometric Approaches to Differential Equations, P. J. VASSILIOU & I. G. LISLE
16 Industrial Mathematics, G. FULFORD & P. BROADBRIDGE

# Chaos

*A Mathematical Introduction*

JOHN BANKS

VALENTINA DRAGAN

ARTHUR JONES

**CAMBRIDGE**
**UNIVERSITY PRESS**

University Printing House, Cambridge CB2 8BS, United Kingdom

One Liberty Plaza, 20th Floor, New York, NY 10006, USA

477 Williamstown Road, Port Melbourne, VIC 3207, Australia

314-321, 3rd Floor, Plot 3, Splendor Forum, Jasola District Centre, New Delhi-110025, India

79 Anson Road, #06-04/06, Singapore 079906

Cambridge University Press is part of the University of Cambridge.

It furthers the University's mission by disseminating knowledge in the pursuit of education, learning and research at the highest international levels of excellence.

www.cambridge.org
Information on this title: www.cambridge.org/9780521531047



First published 2003
Reprinted 2003

*A catalogue record for this publication is available from the British Library*

*Library of Congress Cataloging in Publication data*

621.502′2′015118 – dc21 2001035269

ISBN 978-0-521-53104-7 Paperback

# CONTENTS

# Preface

The purpose of this book is to provide an introduction to chaos theory at a level suitable for university students majoring in Mathematics. The book should also be of interest to scientists and engineers who are familiar with popular approaches to chaos in dynamical systems and wish to learn more about the underlying mathematics.

The subject of *dynamical systems* was founded towards the end of the nineteenth century by the French mathematician Henri Poincaré. The differential equations in which he was interested arose from the study of planetary motion. To make progress in the study of these equations, Poincaré invented new topological methods for studying their solutions, in place of the traditional methods involving series.

Using the new methods which he had invented, Poincaré discovered that the differential equations admitted solutions of a hitherto unimagined complexity. With this discovery, it was realized for the first time that the differential equations describing natural phenomena could have solutions which behaved in a 'chaotic' way. For most of the twentieth century an understanding of Poincaré's work remained the province of a select group of professional mathematicians because of the difficulty of the underlying mathematics.

A new era for the study of differential equations began when computers became available which could quickly generate numerical approximations to their solutions. As a result, chaotic behaviour was soon observed in the differential equations used as models in many areas of science. For example, in 1963 Edward Lorenz published an example of a 'strange attractor'. He had discovered this attractor as a result of using a computer to find approximate numerical solutions to a system of differential equations which he was using to model the weather.

Chaos theory rose to popularity in the 1990's as a topic for television programs and widely read paperbacks. An important factor in this rise was the advent of computers capable of finding approximations to solutions of differential equations and then producing high quality graphics representations of their chaotic behaviour.

Although computers are a wonderful tool for suggesting the presence of chaotic behaviour, they do not normally provide proofs of its existence in the strict mathematical sense. The possibility of introducing undergraduates to a mathematical approach to chaos theory came with the announcement by the mathematical biologist Robert May in 1976, that even very simple models of population growth could lead to chaotic behaviour. These models were *difference equations* in one dimension, as distinct from differential equations in which chaos only appears in dimension three or more.

In this book we use difference equations in one dimension to introduce basic ideas and results of Chaotic Dynamical Systems. Although there are several competing definitions of chaos, we concentrate here on the one given by Robert Devaney, which avoids the use of measure theory and uses only elementary notions from analysis.

An important feature of our approach is the use it makes of the graphs of the composites of the iterator function. We provide rigorous justification for what one sees on the computer screen and use this as the basis on which to build the mathematical theory. This enhances students' intuition for chaos and helps to build their confidence.

The resulting approach is very geometrical in its motivation, both for definitions and for proofs of results. Key ingredients of chaotic mappings — sensitivity to initial conditions, transitivity, and denseness of periodic points — all admit simple interpretations in terms of the graphs of the higher iterates of the original mapping. Although the Schwarzian derivative is one of the key tools used in the study of one-dimensional chaos, its introduction has been poorly motivated in the past. In our approach it arises naturally from the graphs.

The book began as a set of notes for a course that evolved over a number of years with the participation of mathematics students at La Trobe University. Chapters were trialled with students and in the light of their comments on the notes and their attempts at problem solving, chapters were reorganised and rewritten, often many times.

Most students who now take the course do so in their third year. However participants have included first, second, fourth year and graduate students. So although the course is aimed squarely at a strong second to average third year student, it is inherently flexible and is suitable for very strong first year students, as well as more advanced students wanting an introduction to chaotic dynamical systems. (Worked solutions to many of the problems in this book are available to *bona fide* teachers: please contact solution@cambridge.org.)

The early chapters of the book should be accessible to anyone who has completed a course on one-variable calculus which gave some attention to basic concepts (sets; properties of numbers; mathematical induction; functions as entities in their own right, with a clear distinction between the function and its values; composite functions).

Later chapters assume that the students have been exposed to a careful treatment of limits and continuity and have acquired some familiarity with the logic involved in writing simple mathematical proofs. Thus we assume the reader understands the significance of *if ... then* statements, and we use the logical symbols for quantifiers when explaining the meaning of *sensitive dependence.* In our experience, anything less than this means that the student has little chance of successfully completeing the course on which the book is based.

We have found that students often lack the mathematical maturity required to write or understand simple proofs. It helps if students write down where the proof is heading, what they have to prove and how they might be able to prove it. Because this is not part of the formal proof, we indicate this exploration by separating it from the proof proper by using a box which looks like this:

(Include here what must be proved, etc)

In retrospect, there are a number of factors which have made chaos theory a very attractive topic for inclusion in our undergraduate curriculum. For example:

- *It is easy to motivate in terms of 'real world' applications.* The sorts of applications to which students are traditionally exposed in mathematics departments require a good background in physics — a background which many of our students do not have these days. By way of contrast, the most widely studied example in chaos theory comes from an easily understood model of population growth. Moreover, well-produced books, articles and videos are readily available, which explain the relevance of chaos to many areas of science.
- *Some of the most basic results have been discovered within the last couple of decades.* One of the perennial problems of teaching undergraduate mathematics has been that undergraduates rarely reach a level where they can grasp contemporary research problems. The field of Chaotic Dynamics is sufficiently new to admit research questions of an elementary nature which are within their grasp.
- *It uses many of the mathematical concepts and techniques from other parts of undergraduate mathematics.* This helps to give coherence to the undergraduate mathematics course as a whole and gives students the satisfaction of being able to use knowledge obtained previously. In this way it can contribute to breaking down the rigid compartmentalization of mathematical knowledge to which students (and others) often succumb.

As a result we believe that a course on chaotic dynamics, conducted at an appropriate level for our students, has a very positive effect on their understanding of mathematics and their attitude towards it.

**Acknowledgements.** The course on which the book was based began as a series of informal lunch-time working seminars run by John Banks. With the aid of two grants[1], we were able to purchase computing and video equipment and to employ two professional assistants. This allowed us to produce typeset notes for the course, and to video tape the classes and analyse them for signs of student difficulties. The course team included John Banks, Valentina Dragan, Gary Davis, Arthur Jones and Catherine Pearn, with an advisory committee consisting of Jeff Brooks, Grant Cairns, Peter Stacey and Bill Jelks. Their combined input was valuable in producing a course that was accessible and challenging to all students.

We are grateful to our colleagues: Ed Smith for initial encouragement, Robert Johnston, Margaret McIntire, Peter Stacey, Marcel Jackson, Peter Fox, Claire Edwards and Russell Rimmer for help with proof reading, Trevor Clark of the CSIRO for introducing us to the TEXTURES$^{TM}$ package for doing TeX on the Macintosh, and Brian Davey for instructing us in some of the finer points in the use of TeX. Finally, we wish to thank all our students who participated in the course during its developmental stage.

[1] *A Quality Teaching Support Grant* from the Academic Development Unit, La Trobe University, and a *National Teaching Development Grant* from the Committee for The Advancement of University Teaching, Canberra.

# 1

# MAKING PREDICTIONS

*When ancient societies wished to discover what the future held for them they consulted their soothsayers. The methods by which the soothsayers made their predictions now appear to us as quite strange: observing the entrails of animals, noting the position of the planets or viewing the gleam of sacred stones.*

*Nowadays we prefer predictions to be based on scientific theories. Most branches of science embody assumptions (or laws) which can be expressed as mathematical equations. Predictions are made by solving the mathematical equations and then interpreting their solutions in terms of the original scientific problem. In this chapter we illustrate these ideas by drawing on just one area of science: the theory of population growth. This area is sufficiently familiar that its basic assumptions can be understood easily. At the same time, the equations to which it leads can have solutions with extremely complicated behaviour patterns, leading to chaos.*

*The chaotic behaviour of the solutions has far reaching implications for the future of scientific endeavour: for many scientific experiments, accurate predictions of the long term outcomes may not be possible.*

## 1.1 MATHEMATICAL MODELS

In most scientific theories the assumptions (or laws), and the equations to which they lead, do not represent the original problem with complete accuracy. Hence the equations (and the assumptions) are only a *model* whose purpose is to capture the essential features of the original problem while ignoring incidental details.

Predictions about the outcome of the problem are then made by solving the equations. If the predictions of the theory do not agree with the observed outcomes to the desired accuracy, the assumptions of the model are modified to bring them more closely into line with reality, new predictions are made and the process repeated.

### Population growth

Many different mathematical models have been constructed for population growth and they will be our main concern in this chapter. Each of these models is an equation which expresses the rate at which the population is growing in terms of the size of the population.

From the mathematical model we hope to be able to determine the number $N$ of individuals in the population at any later time $t$. Thus we are assuming that $N$ is a function of $t$ and we would like to be able to find this function as explicitly as possible. A simple formula expressing $N$ in terms of $t$ would be the ideal solution, but if this is not possible we would at least like to have a graph showing how $N$ behaves as $t$ increases.

Since a population grows by successive addition of individuals, $N$ must be a discontinuous function of $t$. Between two consecutive times at which individuals are added, the population $N$ will remain constant. Hence the graph of $N$ against $t$ must be that of a step function.

Figure 1.1.1 shows graphs of two step functions. The values of step function (b) have smaller jumps than those of step function (a). Although step functions are discontinuous, in practice they can approximate continuous functions very closely. It is not hard to imagine that if all of the jumps were small enough then the graph of the step function would be indistinguishable, for practical purposes, from that of a continuous function.

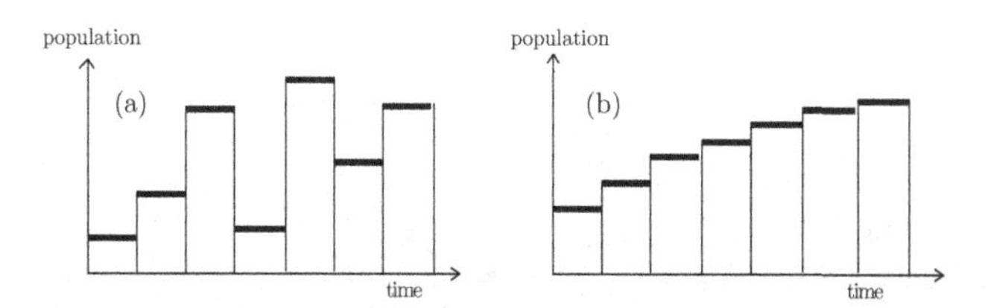


**Fig. 1.1.1** Two step functions (a) and (b). In each case, the population stays constant during a time-interval, and then jumps to a new constant value.

These considerations lead us to consider two main types of models for population growth: *continuous* and *discrete*. We now show how to recognize each type.

## Continuous models

The assumption underlying these models is that *the size of a population varies continuously with time.* As noted above, this assumption can never be strictly true. It is, nevertheless, a reasonable approximation for large populations with no preferred breeding season: for example, human populations or large populations of yeast cells. In such cases the addition of a few individuals will make so little difference to the overall population that an illusion of continuous variation will be produced.

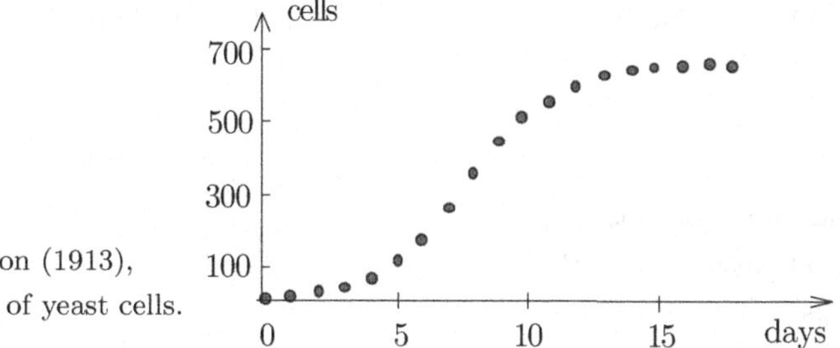


**Fig. 1.1.2**
Data from Carlson (1913), for a population of yeast cells.

In Figure 1.1.2 the observed values for the number of yeast cells in a growing population are plotted. Since this is a population for which a continuous growth model is appropriate, we get the population–time graph by drawing a continuous curve through these points.

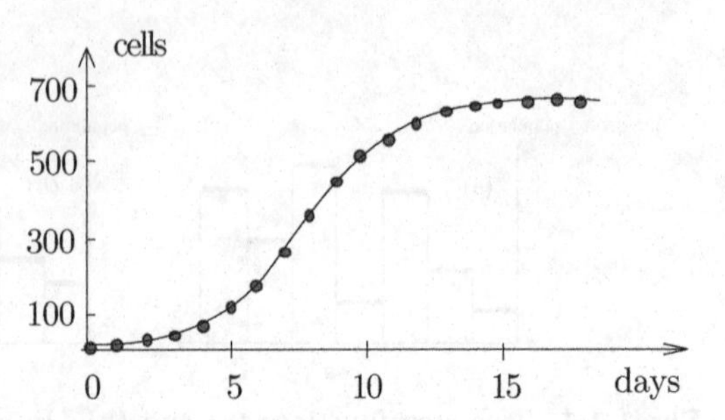


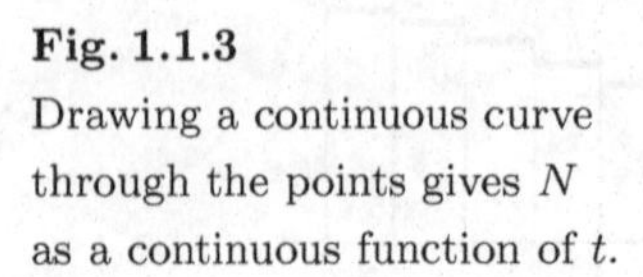

**Fig. 1.1.3**
Drawing a continuous curve through the points gives $N$ as a continuous function of $t$.

An elongated–S (or *sigmoidal*) curve , as shown in Figure 1.1.3, is often associated with a continuously growing population in an environment with only limited food supply. A continuous mathematical model – one which predicts the sigmoidal shape of the curve – will be discussed in the next section.

## Discrete models

These models are appropriate for populations which have specific breeding seasons. Because large numbers of the population all breed at the same time there is a perceptible jump in the size of the population at the end of each breeding season. Thus the assumption underlying these models is that *the size of a population changes abruptly at equally spaced times.* Many types of insects breed in this way, at equally spaced intervals of time.

*Graphs:* For discrete models there are various ways of representing population against time on a graph. One way is as a step function.

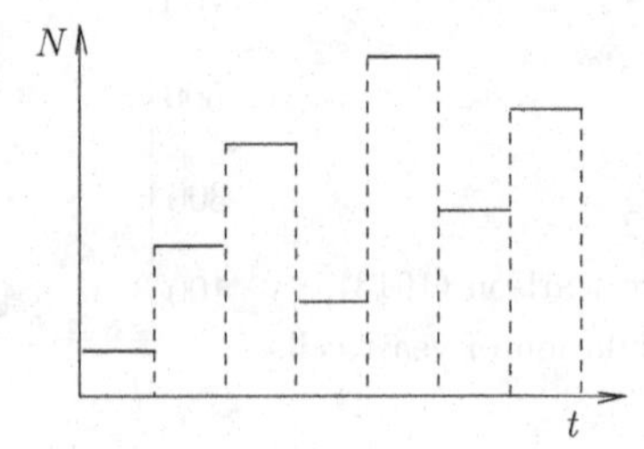


**Fig. 1.1.4**

The breeding seasons are equally spaced, hence the steps have equal width.

Note that there is no change in population between consecutive breeding seasons. It is thus appropriate to measure the size of the population at the end of each breeding season.

This allows us to regard the time $t$ as being a non-negative integer $0, 1, 2, \ldots$ . Hence the graph of population against time consists of discrete points corresponding to these measurements as illustrated in Figure 1.1.5.

**Fig. 1.1.5**
In a discrete model, we regard the time $t$ as a non-negative integer. Hence the graph of the population $N$ against the time $t$ consists of a sequence of isolated points.

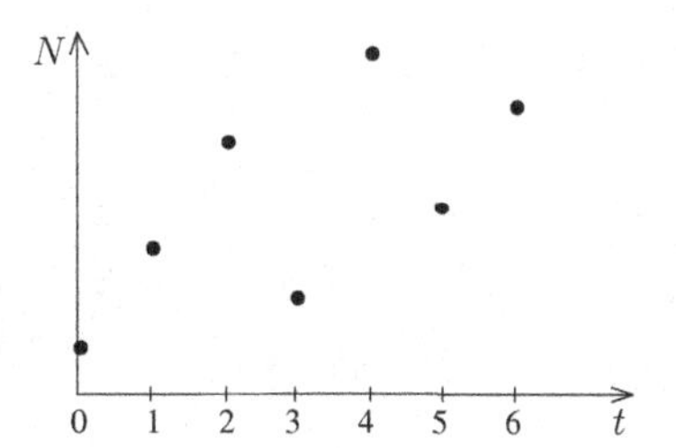


A graph which consists of discrete points, as in Figure 1.1.5, is a theoretically adequate way to represent the data for a discrete model. In practice, however, it is customary to join the dots by line segments to make the pattern of the dots more apparent. Joining the dots we obtain the graph in Figure 1.1.6.

**Fig. 1.1.6**
The line segments joining the dots are not part of the graph. They are there to help us see how $N$ varies from one generation to the next.

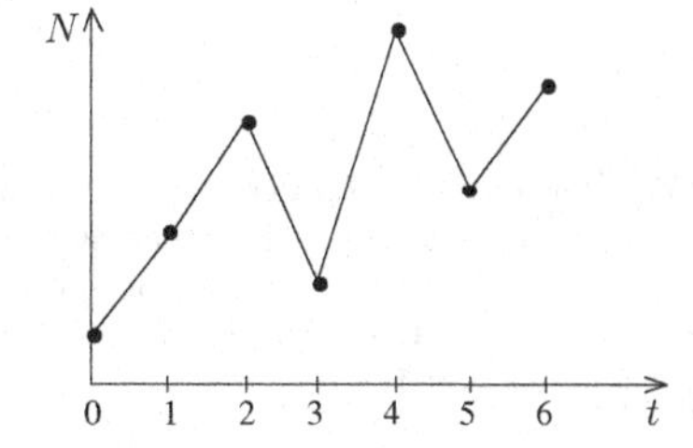


## Growth patterns

The growth curves for continuous models are typically predictable sigmoid curves whereas for discrete models there is a variety of possible growth patterns. They include not only monotonic behaviour but also oscillatory and chaotic behaviour.

Some typical growth patterns for weevils – for which discrete models are appropriate – are shown in Figure 1.1.7.

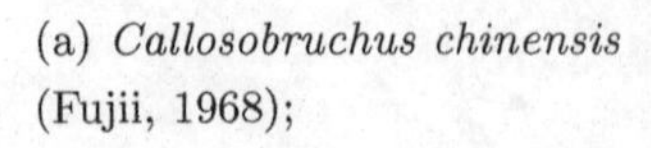

(a) *Callosobruchus chinensis* (Fujii, 1968);

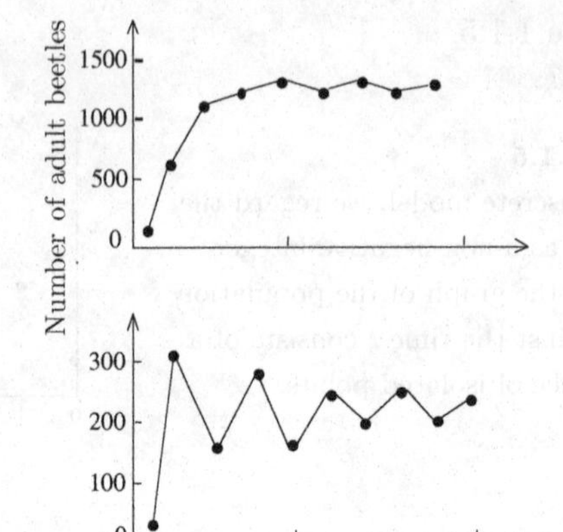


(b) *Callosobruchus maculatus* (Utida, 1967);

(c) *Callosobruchus maculatus* (Fujii, 1968).

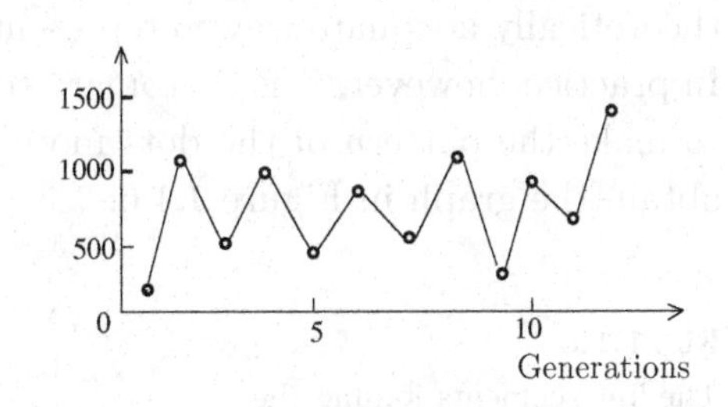


**Fig. 1.1.7** Population changes in laboratory cultures of three different strains of weevils. (Note that the axis which we normally label with the time $t$ is here labelled *generations*. You can think of the first generation as occupying the time interval $0 \leq t \leq 1$. Then think of the population during generation 1 as the population when $t = 0$.)

## Exercises 1.1

1.1.1. Give examples of populations in each category:

(a) a continuous model is appropiate,

(b) a discrete model is appropriate.

1.1.2. For each of the graphs in Figure 1.1.7 :

(a) for how many generations are the number of weevils plotted?

(b) what is the number of weevils during the fifth generation?

(c) what is the population when the time $t = 3$ ?

## 1.2 CONTINUOUS GROWTH MODELS

This section contains two models for the continuous growth of a population. The first is a very simple model in which the effect of limited food and space on the growth are ignored. In the second model, these limitations are taken into account.

In each case we let $N$ denote the number of individuals in the population at time $t$. As we are going to use a differential equation as the model, we ignore the fact that $N$, the number of individuals in the population, must be a whole number and we suppose instead that $N$ is some differentiable function of the time $t$. In each model the derivative

$$\frac{dN}{dt}$$

is *the rate of increase of the population at time* $t$. Hence the ratio

$$\frac{1}{N}\frac{dN}{dt}$$

where $N > 0$ is the *rate of increase of the population per individual.* It is, on average, the number of offspring which each individual produces per unit time and is therefore called the *individual reproduction rate* . If deaths, as well as births, are to be considered then the above ratio is equal to $\{$*birth rate*$\}$ – $\{$*death rate*$\}$ per individual. It can then be negative, but not less than $-1$.

### Unlimited growth model

The simplest model of continuous growth is the assumption that, for a given type of population in a specified environment, *the individual reproduction rate remains constant.* This assumption can be written as a mathematical equation

$$\frac{1}{N}\frac{dN}{dt} = r,$$

where $r$ is a constant, and hence as

$$\frac{dN}{dt} = rN. \tag{1}$$

Thus we have derived a differential equation, which is the mathematical expression of our modelling assumption for population growth[1]. The differential equation has the solution

$$N_t = N_0 \exp(rt). \tag{2}$$

We assume enough data is available to determine the constant $r$. The solution then enables us to predict the size $N_t$ of the population at any time $t$ from its *initial value* $N_0$. For $r > 0$ the population grows exponentially, as indicated in the graph of the solution in Figure 1.2.1.

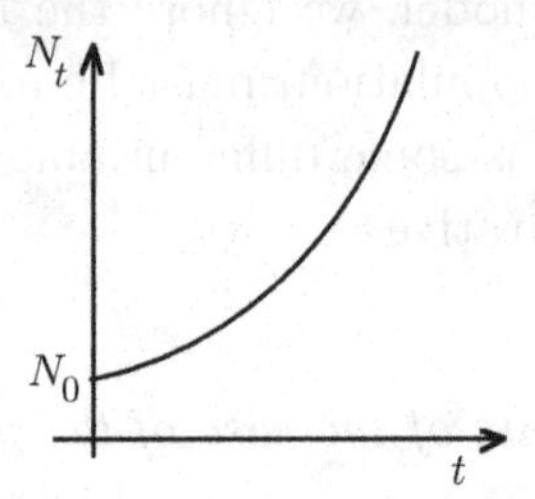


**Fig. 1.2.1**
The unlimited growth which the differential equation (1) predicts.

Thomas Malthus (1766–1834) wrote extensively on the tendency of human populations to grow exponentially. For this reason, the above model is often referred to as the *Malthusian* model. ■

### Limited growth model

For exponential growth to continue indefinitely it is necessary to assume unlimited space and unlimited food supply. In real life, however, population growth is restricted by food, space and other necessities of living. The simplest and most familiar model taking this into account will now be described. It was first introduced in 1844 by Pierre-François Verhulst (1804–1849).

In this model it is assumed that, due to the limitations of the environment, the population has a maximum sustainable size $K$ and

*when the size of the population approaches* $K$*,*
*the individual reproduction rate approaches* 0.

It is also assumed that

*when the size of the population is close to* 0*,*
*the individual reproduction rate is close to a number* $r > 0$.

---

[1] The differential equation (1), unlike the modelling assumption, still makes sense when $N = 0$. Putting $N_t = 0$ for all $t$ gives a *constant* solution of (2).

The simplest assumption for the individual reproduction rate consistent with these two assumptions is that it equals a linear function of $N$ which drops from $r$ to 0 as $N$ increases from 0 to $K$. Hence

$$\frac{1}{N}\frac{dN}{dt} = r\left(1 - \frac{N}{K}\right),$$

and so

$$\frac{dN}{dt} = rN\left(1 - \frac{N}{K}\right). \tag{3}$$

Here $K$ is called the *carrying capacity* of the environment and $r$ is called the *intrinsic reproduction rate.* The differential equation (3) expresses Verhulst's model of population growth mathematically. It is called the *logistic equation.* Note that the differential equation has the constant solutions $N_t = 0$ and $N_t = K$. This equation also has the solution

$$N_t = \frac{K}{1 + ((K/N_0) - 1)e^{-rt}}. \tag{4}$$

where $N_0 > 0$ denotes the initial population, when $t = 0$. We assume enough data is available to determine the constants $r$ and $K$. From the solution (4) we can then predict the size of the population at any time $t$ provided that we are given its initial value $N_0$. Note that as the time $t$ becomes indefinitely large, the population approaches the carrying capacity $K$ and the growth rate approaches zero. The graph of a typical solution is given in Figure 1.2.2.

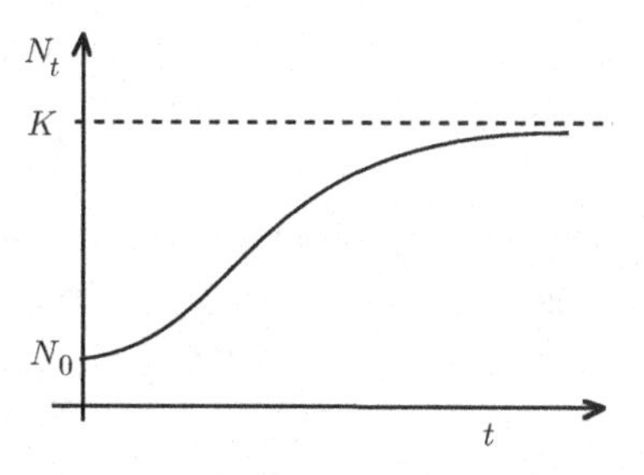


**Fig. 1.2.2**
The limited growth predicted by the differential equation (3).

How well does the logistic model predict the growth behaviour of actual populations? For example, how well does it predict the growth of a population of yeast cells given in Figure 1.1.2 ? According to [Em][2] it is a reasonably good predictor for bacteria, yeast, and

[2] References are given at the end of each chapter.

protozoans. For laboratory populations of water fleas, *Daphnia*, fruit flies, *Drosophila* and sheep it gives a fair fit to data.

In both of our models, whether for limited or unlimited growth, the solutions of the differential equations are given by simple formulae which can be used to make the predictions. Because of this, we say that the solutions can be expressed in *closed form.*

## Exercises 1.2

1.2.1. This exercise refers to the differential equation, with $r$ constant,

$$\frac{dN}{dt} = rN.$$

(a) Show, by separatiing variables, that for $N \neq 0$ the solution is

$$N_t = N_0 \exp(rt).$$

(b) Show the general shape of the graph of a solution in the case $r < 0$ and $N_0 > 0$.

(c) What would you say is happening to the population being modelled by this differential equation when $r < 0$?

1.2.2. This exercise refers to the following differential equation:

$$\frac{dN}{dt} = rN\left(1 - \frac{N}{K}\right) \qquad (r \text{ and } k \text{ constant}).$$

(a) Show, by separating variables, that for $N \neq 0, K$ the solution is

$$N_t = \frac{K}{1 + ((K/N_0) - 1)e^{-rt}}.$$

(b) What does the logistic equation predict when the initial size of the population is greater than the carrying capacity?

1.2.3. For each of the differential equations in the above exercises, find all the constant solutions and discuss their biological interpretation.

## 1.3 DISCRETE GROWTH MODELS

This section is mainly about discrete versions of the two continuous models considered in the previous section.

The basic change is to consider the population only at a set of times which are equally spaced, say at unit time apart. Hence we regard $N_t$, the population at time $t$, as being defined only when $t$ is restricted to the values $0, 1, 2, \ldots$ .

In place of the derivative, we now have the difference $N_{t+1} - N_t$. This represents *the rate of increase in population during the time interval from $t$ to $t+1$.* Hence

$$\frac{1}{N_t}\left(N_{t+1} - N_t\right)$$

is the *the rate of increase of population per individual* during the given time interval.

### Unlimited growth model

The simplest model of discrete growth assumes that *the individual reproduction rate is constant*; that is

$$\frac{1}{N_t}\left(N_{t+1} - N_t\right) = r$$

where $r$ is constant, so $N_{t+1} = N_t + rN_t$ and hence

$$N_{t+1} = N_t(1+r). \tag{5}$$

This is an example of a *difference equation.* Since (5) is to hold for $t = 0, 1, 2, 3, \ldots$, it is equivalent to the infinitely many equations

$$\begin{aligned} N_1 &= N_0(1+r) \\ N_2 &= N_1(1+r) \\ N_3 &= N_2(1+r) \\ &\vdots \end{aligned} \tag{6}$$

A *solution* of the difference equation is an infinite sequence of numbers $N_0, N_1, N_2, N_3, \ldots$ which satisfies the equations (6).

A way to work out a solution is to pick a starting value $N_0$ and then use the first of the equations (6) to calculate $N_1$. We repeat this process: use the second of the equations (6) to find $N_2$. In this way, calculate successively $N_1$, $N_2$, $N_3$, ... in terms of $N_0$. This process of repeated substitution is called *iteration.*

**1.3.1 Example** *Let the sequence $N_0, N_1, N_2, N_3, \ldots$ be a solution of the difference equation (5). Use iteration to express $N_1, N_2, N_3$ in terms of $N_0$.*

*Solution:* Substituting each of the equations in (6) into the equation which follows it gives in turn

$$\begin{aligned} N_1 &= N_0(1+r) \\ N_2 &= N_1(1+r) = N_0(1+r)(1+r) = N_0(1+r)^2 \\ N_3 &= N_2(1+r) = N_0(1+r)^2(1+r) = N_0(1+r)^3. \end{aligned}$$ ■

The above results suggest more generally that every element $N_t$ of the solution can be obtained from the formula

$$N_t = N_0(1+r)^t. \tag{7}$$

The validity of this general formula can be proved by mathematical induction or, alternatively, by checking that

(a) *it gives the correct initial value* and

(b) *when substituted in the difference equation (5) it makes both sides equal.*

The following example illustrates this method.

**1.3.2 Example** *Show, by substitution, that the formula (7) gives the solution of the difference equation (5) with initial value $N_0$.*

*Solution:* Putting $t = 0$ in (7) gives the initial value $N_0(1+r)^0$, which is correctly equal to $N_0$.

Substituting (7) into the difference equation (5) gives, moreover,

$$\begin{aligned} \text{LHS} &= N_{t+1} = N_0(1+r)^{t+1} \\ \text{RHS} &= N_t(1+r) = N_0(1+r)^t(1+r) = N_0(1+r)^{t+1}. \end{aligned}$$

Thus the two sides are equal. Hence (7) gives the solution of (5) satisfying the required initial condition. ■

We say that (5) has a *closed form* solution since we have been able to find a simple formula for the solution. It is only in exceptional cases, however, that we shall be able to do this.

### Limited growth model

The assumptions here are analogous to those in the limited growth model for continuous growth in the previous section. Hence

$$\frac{1}{N_t}(N_{t+1} - N_t) = r\left(1 - \frac{N_t}{K}\right)$$

and so

$$N_{t+1} = N_t + rN_t\left(1 - \frac{N_t}{K}\right). \tag{8}$$

This difference equation is our limited growth model for discrete growth.

### Other models

Other models are given by the difference equations

$$N_{t+1} = \lambda N_t(1 + aN_t)^{-\beta} \tag{9}$$

and

$$N_{t+1} = \lambda N_t \exp(-\alpha N_t). \tag{10}$$

In both (9) and (10) $\lambda$ is the growth rate when the population is small and $a$, $\alpha$ and $\beta$ are constants.

It was easy to show that the difference equation (5) has a closed form solution. Deciding whether (8), (9) and (10) have closed form solutions is not so easy. In principle, however, we can solve them using iteration. Their solutions have a variety of types of behaviour: monotonic, cyclic, damped oscillatory and chaotic.

The growth model (8) is of limited interest to biologists. It is hard, in fact, to find population data which can be predicted by it. It has the unrealistic feature, moreover, that if $N_t$ is large enough, then $N_{t+1}$ will be negative. In the last two models, however, by choosing the parameters we can get solutions to fit any one of the three sets of observations given in Figure 1.1.7. Hence these models appear to be more realistic than (8). Each of the models in (8) and (9) may be derived from a particular set of biological assumptions.

### Scaling and parameters

The two constants $r$ and $K$ which appear in the difference equation (8) are called *parameters.* The idea behind this terminology is that (8) is essentially not just *one* difference equation but a *family* of such equations — one equation for each choice of $r$ and $K$. This gives us flexibility in modelling, since we can choose these numbers in a way which best fits the data for a particular population in a given environment.

A standard technique in mathematical modelling is rescaling so as to lump together as many parameters as possible into a single parameter. We shall illustrate this by showing how the two parameters in (8) can be replaced by one parameter $\mu$.

To achieve this, the trick is to introduce a 'scaled' population $x_t$ in place of $N_t$ by putting

$$x_t = \frac{r}{(r+1)K} N_t \qquad (t = 0, 1, 2, \ldots).$$

Solving this equation for $N_t$ and then substituting in (8) gives the difference equation

$$x_{t+1} = \mu x_t (1 - x_t) \tag{11}$$

where $\mu = 1 + r$. This gives the scaled population at the end of the $(t+1)^{\text{th}}$ breeding season in terms of that at the end of the $t^{\text{th}}$ breeding season. The difference equation (11) is called the *discrete logistic equation* (with parameter $\mu$.)

## Exercises 1.3

1.3.1. A solution $x_0, x_1, x_2, x_3, \ldots$ of a difference equation is said to be *constant* if all of the $x$'s are the same. Find all the constant solutions of the difference equation

$$x_{t+1} = x_t(1 - x_t) \qquad (t = 0, 1, 2, 3, \ldots) .$$

1.3.2. Let $\mu > 0$. Repeat Exercise 1 for the difference equation

$$x_{t+1} = \mu x_t(1 - x_t).$$

1.3.3. For each of the difference equations given in the text, find all the constant solutions.

1.3.4. This exercise refers to the solution of the difference equation

$$x_{t+1} = (x_t)^2$$

which satisfies the initial condition $x_0 = 2$.

(a) Use iteration to find the next three elements $x_1, x_2, x_3$ of the solution.

(b) Guess a general formula for $x_t$. Check your answer by substituting it in the difference equation.

(c) What happens to $x_t$ as $t$ approaches $\infty$?

1.3.5. Repeat Exercise 4, but this time use the initial condition $x_0 = 1$. What do you notice about the solution?

1.3.6. This exercise refers to the solution of the difference equation

$$x_{t+1} = \sqrt{x_t}$$

which satisfies the initial condition $x_0 = 64$.

(a) Use iteration to find the next three elements $x_1, x_2, x_3$ of the solution.

(b) Guess a general formula for $x_t$, valid for $t = 0, 1, 2, 3, \ldots$ . Check your answer by substituting it in the difference equation.

(c) What happens to $x_t$ as $t$ approaches $\infty$?

1.3.7. Repeat Exercise 6, but this time use the initial value $x_0 = 1$. What do you notice about the solution?

1.3.8. We can define the sum $s_n = 1+2+3+\cdots+(n-1)+n$ of the first $n$ positive integers recursively by putting

$$s_1 = 1$$
$$s_{n+1} = s_n + (n+1) \qquad \text{(for } n = 1, 2, 3, \dots)$$

(a) Check that the solution to this difference equation which satisfies the initial condition is obtained by putting

$$s_n = \frac{n(n+1)}{2}$$

(b) For the purpose of calculating, say $s_{1000}$, which do you think is the more convenient:

- iteration using the difference equation (1), or
- using the quadratic formula for the solution from part (a) ?

1.3.9. Derive the scaled form (11) of the difference equation (8).

(a) in the special case where $r = 1$ and $K = 2$,

(b) in the general case where there is no restriction on these parameters.

[Hint. In each case first write the substitution so it expresses $N_t$ in terms of $x_t$.]

1.3.10. By using a suitable substitution, show how to reduce the differential equation

$$\frac{dN}{dt} = rN\left(1 - \frac{N}{K}\right)$$

to a differential equation involving just one parameter.

[Hint. This differential equation is the continuous analogue of the difference equation (8) in the text.]

## 1.4 NUMERICAL SOLUTIONS

In this section we look briefly at the behaviour of solutions of the discrete logistic equation

$$x_{t+1} = \mu x_t(1 - x_t) \tag{12}$$

As we mentioned in the previous section, iteration is used to work out solutions of discrete equations. It is known that (12) has a closed form solution when $\mu = 2$ and $\mu = 4$. For any other values of $\mu$, we do not know whether (12) has a closed form solution. We use the next example and graphs of solutions to illustrate the behaviour of the solutions of (12) for various values of $\mu$.

**1.4.1 Example** *Let $\mu = 1$. Use iteration to find the first four elements of the solution $(x_0, x_1, x_2, x_3, \ldots)$ of (12) which satisfies the initial condition $x_0 = 2$. Guess what happens to $x_t$ when $t$ is large.*

*Solution:* Replacing $\mu = 1$ in the formula and $x_0 = 2$ we get

$$\begin{aligned} x_1 &= x_0(1 - x_0) = -2 \\ x_2 &= x_1(1 - x_1) = -6 \\ x_3 &= x_2(1 - x_2) = -42. \end{aligned}$$

We guess that as $t$ increases $x_t$ becomes increasingly large and negative, so we guess that $x_t$ approaches $-\infty$ as $t$ approaches $\infty$. ∎

More generally, for initial values outside the interval $[0, 1]$ the solutions very quickly approach minus infinity; hence we are normally interested only in solutions with initial values in $[0, 1]$.

### Graphs of solutions

The graph of the solution of a difference equation is obtained by plotting *the value of $x_t$ against $t$*. Typical graphs of solutions of a difference equation are shown in Figure 1.4.1. If enough points are plotted on the graph, we can sometimes guess the behaviour of the solution from the pattern of the dots.

In Figure 1.4.1 we plot the solution of the discrete logistic equation (12) with initial value $x_0 = .85$, for different values of $\mu$. As $\mu$ increases we observe different types of behaviour.

(a) For $\mu = 2.5$ the solution tends toward a single value. (For $1 \leq \mu \leq 3$, the results are similar. The limiting value increases gradually as $\mu$ increases.)

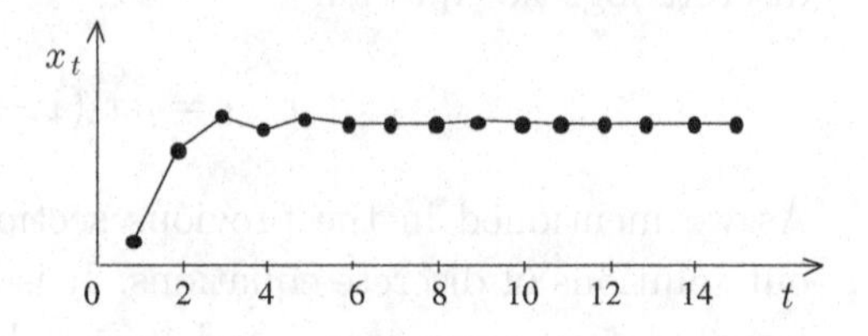


(b) For $\mu = 3.1$ the solution eventually settles down to oscillate between two specific values.

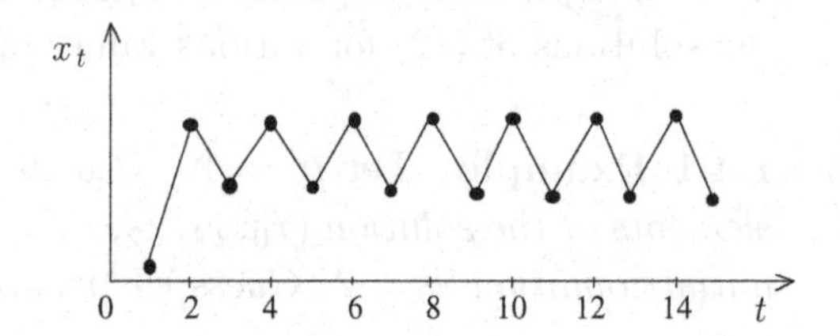


(c) For $\mu = 3.5$ the solution eventually settles down to oscillate between four specific values.

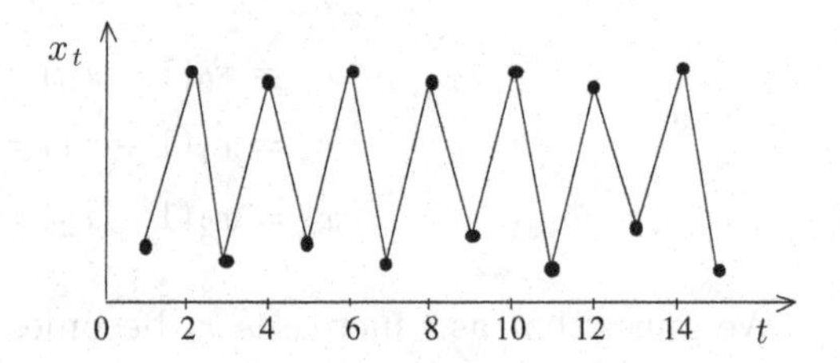


(d) For $\mu = 4$ the solution appears to bounce around in a random fashion. This looks like chaos!

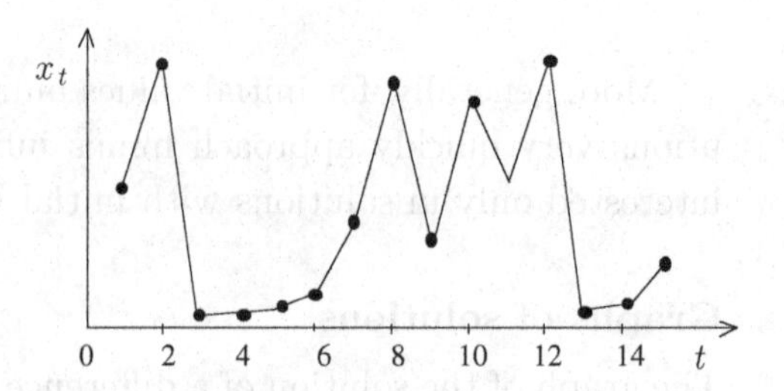


**Fig. 1.4.1** The above results suggest that the solutions of the difference equation become more complicated as $\mu$ increases. In particular, the solution with initial condition $x_0 = .85$ evolves from a constant (when $\mu = 2.5$) to an oscillation between two specific values (when $\mu = 3.1$) to an oscillation between four specific values (when $\mu = 3.5$) and finally to chaos (when $\mu = 4$.)

## Exercises 1.4

1.4.1.

(a) Use iteration to find the solution of the difference equation

$$x_{t+1} = -x_t \qquad (t = 0, 1, 2, 3, \dots)$$

with initial value $x_0 = 1$. Sketch the graph of the solution and describe its behaviour verbally.

(b) Repeat part (a) when the initial condition is $x_0 = -1$.

1.4.2. Use iteration to find the solution of the difference equation

$$x_{t+1} = 4x_t(1 - x_t) \tag{X}$$

which satisfies the initial condition $x_0 = \frac{1}{2}$. Sketch the graph of the solution and describe its behaviour verbally.

1.4.3. Consider the difference equation

$$y_{t+1} = 4y_t(1 - 2y_t). \tag{Y}$$

(a) Show that if you make the 'change of variables' $y_t = \frac{1}{2}x_t$ then you get the difference equation (X) of Exercise 2.

(b) Which change of variables, when applied to the difference equation (X), gives the difference equation (Y) ?

(c) Use the solution you found for the difference equation (X) and the change of variables from part (a) to deduce a solution of (Y).

1.4.4. Find $c \in \mathbb{R}$ such that, under the change of variables $x_t = y_t + c$, the difference equation

$$x_{t+1} = 2x_t + 1$$

reduces to the difference equation

$$y_{t+1} = 2y_t.$$

## 1.5 DYNAMICAL SYSTEMS

In this chapter we have used models for population growth as an introduction to *dynamical systems:* the population changed with time and we wanted to make long-term predictions about its growth. We were interested in such questions as:

(a) *Does the population become indefinitely large as time increases?*

(b) *Does the population eventually die out ?*

(c) *Does the population increase monotonically or does it oscillate about some value?*

(d) *Are there values of the parameters for which the population grows in a chaotic, unpredictable way?*

To answer these questions, we set up mathematical models describing the rate of growth of the population and then solved, or attempted to solve, the resulting differential or difference equations.

Population growth, however, is only one of many areas in which the idea of a dynamical system arises. Devaney describes dynamical systems as 'the branch of mathematics that attempts to describe processes in motion'. He goes on to say that

> Such processes occur in all branches of science. For example, the motion of the stars and the galaxies in the heavens is a dynamical system, one that has been studied for centuries by thousands of scientists. The ups and downs of the stock market is another system that changes in time, as is the weather throughout the world. The changes chemicals undergo, the rise and fall of populations, and the motion of a simple pendulum are classical examples of dynamical systems in chemistry, biology and physics. Clearly dynamical systems abound [De2].

What is important about 'dynamical systems', of course, is that there is a law which describes their evolution over time. In the models of population growth studied in this chapter, the law was given by a differential equation or a difference equation.

To model real world situations, such as those described by Devaney, the first step often is to find the right differential or difference equation — one whose solutions model the observed evolution. As we show later, in Section 2.2, this is sometimes impossible, and the theory of dynamical systems is then inapplicable.

### Dimension

The dynamical systems considered so far in this chapter have been *one-dimensional* in the sense that the size of the population at time $t$ is a real number $N_t$ and hence can be regarded as a point which moves in time on the real line (a one-dimensional space). These models took account of only one species of population. In real life, however, species do not exist in isolation, but interact with other species.

For example, in the Lotka–Volterra model, two species are considered simultaneously: predator and prey. If $M_t$ and $N_t$ are the numbers of predators and prey at time $t$, then the model for the growth of the two species is the pair of differential equations

$$\frac{dM}{dt} = (a - bN)M$$
$$\frac{dN}{dt} = (-c + dM)N$$

where $a, b, c, d$ are positive constants.

Instead of thinking of $M_t$ and $N_t$ separately, we can combine them into a single pair of real numbers $(M_t, N_t)$. Geometrically, this is a point which moves, as time increases, in the *plane* $\mathbb{R}^2$. For this reason we regard the predator-prey model as an example of a *two-dimensional* dynamical system.

An example of a discrete two-dimensional model is provided by a model for the spread of certain types of infectious diseases (like whooping cough or measles) which is described in the article [AM]. If $C_t$ denotes the number of cases and $S_t$ the number of susceptibles at time $t$, then the rate at which the disease spreads is modelled by the pair of difference equations

$$C_{t+1} = fC_tS_t$$
$$S_{t+1} = S_t + B - fC_tS_t$$

where $f$ and $B$ are positive constants.

### Continuous versus discrete

Part of the folklore is that a continuous dynamical system can exhibit chaotic behaviour only if it has dimension greater than two.

On the other hand, discrete dynamical systems can show chaotic behaviour even in dimension one. The realization of this fact by the Australian mathematical biologist Robert M. May in 1976 was an important factor in the recent rise to prominence of chaotic dynamical systems. Because of this, the basic ideas associated with chaotic behaviour can be introduced with much simpler mathematical tools than was previously thought possible.

The rest of this book will deal exclusively with discrete dynamical systems. It should be realized, however, that there are close analogies between discrete and continuous dynamical systems.

For example the famous Restricted Problem of Three Bodies, from astronomy, is a continuous dynamical system. Towards the end of the nineteenth century Henri Poincaré, the founder of the modern theory of dynamical systems, solved this problem by first reducing it to the study of a discrete dynamical system, in two dimensions.

Again, numerical approximations to the solutions of differential equations are usually obtained by first approximating the differential equation by a difference equation. Thus some knowledge of discrete dynamical systems is useful even for those who are primarily interested in continuous ones.

---

## Additional reading for Chapter 1

A brief discussion of discrete models for population growth is given in Section 1.6 of [De2]. On pages 1-7 of [De1] both the continuous and the discrete logistic models for population growth are discussed briefly, and then used to illustrate the idea of a dynamical system. [PMJPSY] Section 5-8 discusses discrete population growth models. [PJS], pages 42-48 concentrates on the discrete logistic model for population growth and uses it to introduce the idea of chaotic behaviour.

Both [Ste], Chapter 13, and [Ma2] give an overview of the rich variety of dynamical behaviour (including chaotic behaviour) which can be exhibited by the solutions of simple models of population growth.

[Gl], pages 57–80, gives some interesting recent history concerning population models and the part which they played in the development of Chaotic Dynamics.

[Ma1] is the article in which Robert May announced that simple biological models could exhibit complicated dynamical behaviour. The first page or so should be understandable on the basis of what we have covered in the present chapter, and the rest of the article on the basis of our subsequent chapters.

[HLM] contains tables of data for a large number of populations. The concluding discussion, as to how many of the populations showed chaotic growth patterns, is likely to be of general interest.

The experimental observations reported in Section 1.2 come from [Fu1], [Fu2] and [Ut]. See also [HLM]. Very convincing experimental evidence of chaotic fluctuation in insect populations is also given in [CCDD], and [Ka] contains an interesting discussion of the ecological implications of these ideas.

For an easy to follow pictorial explanation of some of the main ideas in the study of continuous models, leading to a discussion of chaos in this context, see [AS] parts 1 and 2.

---

## References


[AM] Roy Anderson and Robert May, "The logic of vaccination", *New Scientist,* (18 Nov 1982), 410-415.

[AS] Ralph Abraham and Christopher Shaw, *Dynamics, The Geometry of Behavior* Aerial Press, Santa Cruz, 1988. In four volumes:

Part 1: Periodic Behaviour

Part 2: Chaotic Behaviour

Part 3: Global Behaviour

Part 4: Bifurcation Behaviour

[Ca] T. Carlson "Über Geschwindigkeit und Grösse der Hefevermehrung" *Würze. Biochem Z.* 57 (1913) 313-334.

[CCDD] R. Costantino, J. Cushing, Brian Dennis and Robert Desharnais, "Experimentally Induced Transitions in the Dynamic Behaviour of Insect Populations", *Nature,* 375 (18 May 1995), 227-230.

[De1] Robert L. Devaney, *An Introduction to Chaotic Dynamical Systems: Second Edition,* Addison-Wesley, Menlo Park California, 1989.

[De2] Robert L. Devaney, *Chaos, Fractals and Dynamics, Computer Experiments in Mathematics,* Addison-Wesley, New York, 1990.

[Em] J.M.Emlen, *Population Biology: The Coevolution of Dynamics and Behaviour,* Macmillan, New York, 1984.

[FFJ] Glenn Fulford, Peter Forrester and Arthur Jones, *Modelling with Differential and Difference Equations.* Cambridge University Press, Cambridge, 1997.

[Fu1] Koichi Fujii, "Studies on Interspecies Competition between the Azuki Bean Weavil, *Callobruchus Chinensis,* and the Southern Cowpea Weevil, *C. Maculatus.* II. Competition under Different Environmental Conditions", *Research on Population Ecology,* IX(1967), 192-200.

[Fu2] Koichi Fujii, "Studies on Interspecies Competition between the Azuki Bean Weavil and the Southern Cowpea Weevil. III. Some Characteristics of Strains of Two Species, *Research on Population Ecology,* X(1968), 87-98.

[Gl] James Gleick, *Chaos: Making a New Science,* Sphere Books, London, 1988.

[HLM] M. P. Hassel, J. H. Lawton and R. M. May, "Patterns of Dynamical Behaviour in Single-Species Populations", *Journal of Animal Ecology,* 451(1976), 471-486.

[Ka] Peter Kareiva,"Predicting and Producing Chaos", *Nature,* 375 (18 May 1995), 189-190.

[Ma1] Robert M. May, "Simple mathematical models with very complicated dynamics", *Nature,* 261 (1976), 459-467.

[Ma2] Robert May, *Theoretical Ecology; Principles and Applications* Blackwell, 1976.

[PMJPSY] Hans-Otto Peitgen, Evan Maletsky, Hartmut Jürgens, Terry Perciante, Dietmar Saupe, and Lee Yunker, *Fractals for the Classroom: Strategic Activities Volume Two,* Springer-Verlag, 1992. (two volumes)

[PJS] Hans-Otto Peitgen, Hartmut Jürgens and Dietmar Saupe, *Chaos and Fractals, New Frontiers of Science*, Springer-Verlag, New York, 1992.

[Ste] Ian Stewart, *Does God Play Dice?* Penguin Books, Middlesex, 1989.

[Ut] Syunro Utida, "Damped Oscillation of Population Density at Equilibrium", *Researches on Population Ecology,* IX(1967), 1-9.

# 2

# MAPPINGS AND ORBITS

*In the previous chapter it was shown how the study of population growth leads to difference equations, such as the logistic equation*

$$x_{n+1} = \mu x_n(1 - x_n)$$

*where $n \in \{0, 1, 2, 3, \dots\}$. The idea of a solution of a difference equation was explained and some possible types of behaviour for solutions were discussed.*

*In this chapter the emphasis shifts from the difference equations to the functions (or mappings) which occur on their right-hand sides. The terminology undergoes a corresponding change: a solution of a difference equation is now regarded as a 'sequence of iterates' of some initial point under a mapping. We regard the iterates as the result of repeated application of the mapping to the initial value and hence refer to the mapping as the 'iterator mapping'.*

*As a result of this change of viewpoint, there are now two different types of graphs to be considered: the graph of the sequence of iterates and the graph of the iterator mapping.*

## 2.1 MAPPINGS

First recall that a difference equation – such as the logistic equation – expresses $x_{n+1}$ as a function of $x_n$ so that

$$x_{n+1} = f(x_n) \tag{1}$$

for some mapping (or function) $f$. The values of $x_n$ and $x_{n+1}$ are usually restricted to lie in some set $S$. We therefore assume that the mapping $f$ has this set $S$ for both domain and codomain and indicate this by writing

$$f : S \to S.$$

We call $f$ the *right-handside mapping* for the difference equation (1).

Conversely, each mapping $f : S \to S$ is the right-hand side mapping for the difference equation (1). Thus there is a one-to-one correspondence between difference equations and mappings.

The set $S$ depends on the particular problem which we are modelling. In the simplest cases it is a subinterval of the real line $\mathbb{R}$. In models involving a pair of unknown quantities, however, $S$ is a subset of the plane $\mathbb{R}^2$. It is sometimes convenient to assume that $S$ is a subset of the complex plane $\mathbb{C}$.

**2.1.1 Example** *Find the right-hand side mapping for the logistic equation with parameter $\mu = 2$. Sketch the graph of this mapping.*

*Solution:* When $\mu = 2$, the logistic equation $x_{n+1} = \mu x_n(1 - x_n)$ is

$$x_{n+1} = f(x_n) \qquad \text{where} \qquad f(x_n) = 2x_n(1 - x_n).$$

Since the population can never be negative, both $x_n$ and $x_{n+1}$ must be non-negative. This will be so if both $x_n \geq 0$ and $x_n \leq 1$, that is, if $x_n \in [0, 1]$. Hence the right-hand side mapping is $f : S \to S$ where

$$f(x) = 2x(1 - x)$$

and $S$ is the interval $[0, 1]$. Figure 2.1.1 shows the graph of $f$.

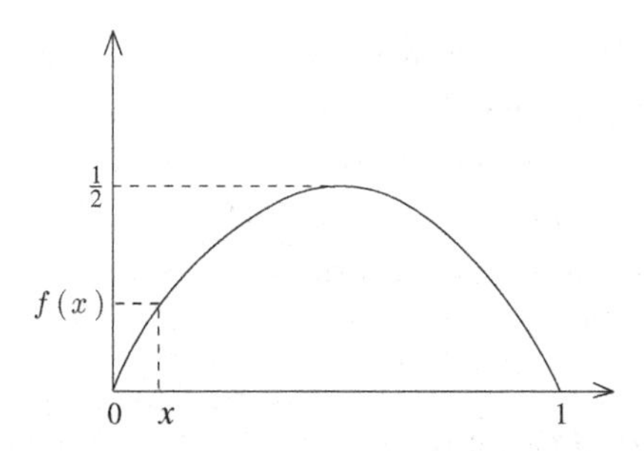


**Fig. 2.1.1** RHS mapping for the logistic equation with $\mu = 2$.

We now begin the process of translating concepts for difference equations into concepts for mappings. Recall that a *solution* of the difference equation (1) is a sequence of elements of $S$,

$$(x_0, x_1, x_2, \dots), \tag{2}$$

such that (1) is true for

$$n = 0, 1, 2, \dots .$$

Writing out the equation (1) for these values of $n$ gives the infinitely many equations

$$\begin{aligned} x_1 &= f(x_0) \\ x_2 &= f(x_1) \\ x_3 &= f(x_2) \\ &\vdots \quad . \end{aligned} \tag{3}$$

Thus the sequence (2) is a solution of the difference equation if and only if *all* of these equations are satisfied.

To solve the equations (3) and hence (1) systematically, we propose to *choose* $x_0 \in S$ *and then use the equations to give successively*

$$x_1, x_2, x_3, \dots .$$

Since $x_0 \in S$ and $f$ maps the set $S$ into itself, a simple inductive argument shows that all of the elements lie in $S$. Hence the proposed procedure can be continued indefinitely to produce a sequence of elements satisfying the difference equation (1). The sequence so obtained is determined uniquely by $x_0$ and we refer to it as the solution of equation (1) which has $x_0$ as its *initial value.*

The procedure by which we calculate the iterates using (3) is called *iteration*. When expressed in terms of mappings – rather than difference equations – the sequence determined by this procedure is called the sequence of *iterates of* $x_0$ *under* $f$. We shall call the mapping $f$, whose repeated application produced the iterates, the *iterator mapping* or just the *iterator*.

**2.1.2 Example** *Let* $f : \mathbb{R} \to \mathbb{R}$ *where* $f(x) = 2x + 1$. *Find the first four iterates of* $0$ *under* $f$ *and guess a formula for the* $n$*th iterate.*

*Solution:* We use the notation as in (3). Hence

$$\begin{aligned} x_0 &= 0, \\ x_1 &= f(0) = 1, \\ x_2 &= f(1) = 3, \\ x_3 &= f(3) = 7, \\ x_4 &= f(7) = 15. \end{aligned}$$

From these results we guess that the $n$th iterate of $x_0$ is $2^n - 1$. ■

**2.1.3 Example** *Use mathematical induction to prove the result guessed in Example 2.1.2.*

*Solution:* We shall prove that, for all $n \in \mathbb{N}_0$,

$$x_n = 2^n - 1. \tag{4}$$

First note that $x_0 = 0 = 2^0 - 1$, and so the formula (4) holds when $n = 0$.
Second let $n \geq 0$ be an integer for which the formula is true. Hence

$$x_{n+1} = f(x_n) = 2x_n + 1 = 2(2^n - 1) + 1 = 2^{n+1} - 1.$$

Thus the formula (4) is valid with $n+1$ in place of $n$. By mathematical induction, the formula is valid for all $n \in \mathbb{N}_0$. ■

In the next example we make a brief digression into the complex plane.

**2.1.4 Example** *Let the set $S$ be the complex plane $\mathbb{C}$, let $i$ be the complex number $\sqrt{-1}$ and let $f : \mathbb{C} \to \mathbb{C}$ where $f(z) = zi$. Find the first four iterates of 1 under $f$ and say what happens to the remaining iterates.*

*Solution:* We call the sequence of iterates $z_0, z_1, z_2, \ldots$, so that

$$\begin{aligned} z_0 &= 1, \\ z_1 &= f(1) = i, \\ z_2 &= f(i) = i^2 = -1, \\ z_3 &= f(-1) = -i, \\ z_4 &= f(-i) = i(-i) = 1, \\ &\vdots \end{aligned}$$

Thus $z_4$ has gone back to $z_0$. Hence repeated application of $f$ to $z_4$ will do exactly the same as it did to $z_0$. Hence the sequence of iterates will consist of the first four numbers $z_0, z_1, z_2, z_3$, repeated indefinitely. ∎

The next example gives a general result about a sequence of iterates which, although very simple, is none the less useful.

**2.1.5 Example** *Show that if $(x_0, x_1, x_2, \ldots)$ is the sequence of iterates of $x_0$ under $f$, then $(x_k, x_{k+1}, x_{k+2}, \ldots)$ is the sequence of iterates of $x_k$ under $f$, for each $k \in \mathbb{N}_0$.*

*Solution:* Let us write the sequence of iterates of $x_0$ under $f$ as

$$(x_0, x_1, x_2, \ldots, x_k, x_{k+1}, x_{k+2}, \ldots).$$

This means that each element (with the exception of $x_0$) is obtained from the previous one by applying the mapping $f$.

Hence each element of the sequence

$$(x_k, x_{k+1}, x_{k+2}, \ldots),$$

(with the exception of $x_k$) is obtained from the previous element by applying $f$. This sequence is therefore the sequence of iterates of $x_k$ under $f$. ∎

## Exercises 2.1

2.1.1. Let $x_n \geq 0$ for $n \in \mathbb{N}_0$. What is the right-hand side mapping for the difference equation $x_{n+1} = 2 + \sqrt{x_n}$ ?

2.1.2. Write down the difference equation which has $f : \mathbb{R} \to \mathbb{R}$ as its right-hand side mapping where $f(x) = x + \cos(x^2)$.

2.1.3. Let $f : [0, 1] \to [0, 1]$ with $f(x) = \frac{x}{x+1}$.

(a) Find the first four iterates of 1 under $f$.

(b) Guess a formula for the $n$th iterate of 1 under $f$ and prove your guess is valid for each $n \in \mathbb{N}_0$. Use mathematical induction.

(c) Sketch the graph of the iterator and the graph of the iterates.

2.1.4. Let $f : \mathbb{R} \to \mathbb{R}$ with $f(x) = x^3$.

(a) Find the third iterate of 2 under $f$.

(b) Guess a formula for the $n$th iterate of 2 under $f$ and prove your guess is valid by using mathematical induction.

2.1.5.

(a) Express Example 2.1.2 as an exercise about the solution of a difference equation with a certain initial condition.

(b) Check that $x_n = 2^n - 1$ satisfies the difference equation and the initial condition you have written down in part (a). Use the method explained in Example 1.3.2.

2.1.6. Prove (as claimed in the text) that if $f : S \to S$ and $x_0 \in S$, then every iterate of $x_0$ under $f$ is also in $S$. Use mathematical induction.

2.1.7. Let $f : \mathbb{C} \to \mathbb{C}$ with

$$f(z) = \left( \frac{-1 + \sqrt{3}i}{2} \right) z.$$

Find the first three iterates of $z_0 = 1$ under $f$ and say what happens to the remaining iterates.

## 2.2 TIME SERIES

In the previous section we started from a mapping $f : S \to S$, which we called the *iterator mapping* and then produced a sequence of *iterates* $(x_0, x_1, x_2, \dots)$ from an *initial value* $x_0 \in S$ by repeatedly applying the mapping $f$.

However, in many applications one wishes to follow the reverse path. The starting point is not an iterator mapping, but a sequence of observations $(x_0, x_1, x_2, \dots)$ of some quantity, made at successive times $t = 0, 1, 2, \dots$.

The sequence of observed values is called a *time series.*[1] The observations could be the daily averages of the stock market at the end of each day's trading, or the year by year estimates of the depletion of the ozone layer. The graphs of the weevil populations shown in Figure 1.1.7 are graphs of time series.

### Reconstructing an iterator mapping

Mathematicians involved in modelling such systems ask of a given time series

$$(x_0, x_1, x_2, \dots, x_n, x_{n+1}, \dots) \tag{5}$$

whether it is random or whether there is some way in which subsequent observations can be predicted from the earlier ones. In particular they are interested to know whether or not there exists an iterator mapping for the time series (5); that is, *whether or not there is a mapping $f : S \to S$ such that, for all $n \in \mathbb{N}_0$,*

$$f(x_n) = x_{n+1}. \tag{6}$$

In the following examples we assume that $f : [0,1] \to [0,1]$ and show various ways in which (6) can be used in the search for an iterator mapping.

[1]The use of the word 'series' in this context is unfortunate since it is being used as a synonym for 'sequence'. In calculus, however, the two words mean two different things and need to be carefully distinguished.

**2.2.1 Example** *Decide whether or not there is an iterator mapping for the sequence*

$$\left(1, \frac{1}{2}, \frac{1}{3}, \frac{1}{4}, \frac{1}{5}, \ldots\right).$$

*Solution:* In this example we make use of the fact that there is a simple formula for the $n$th element of the sequence. We want a mapping $f : [0,1] \to [0,1]$ with $f(x_n) = x_{n+1}$; that is,

$$\begin{aligned} f\left(\frac{1}{n}\right) &= \frac{1}{n+1} \\ &= \frac{\frac{1}{n}}{\frac{1}{n}+1}. \end{aligned}$$

We therefore choose $f(x) = \frac{x}{x+1}$, for all $x \in [0,1]$. The mapping $f$ is then an iterator for the time series. ■

**2.2.2 Example** *Decide whether or not there is an iterator mapping for the sequence*

$$(1, 2, 3, 1, 3, 2, 1, 2, 3, \ldots).$$

*Solution:* In this example $x_0 = 1$ and $x_3 = 1$. Hence if $f$ were an iterator mapping for this sequence then it would have to satisfy both of the conditions

$$f(1) = 2 \quad \text{and} \quad f(1) = 3.$$

But this is impossible since a mapping cannot send a given point in its domain to two different points in its codomain. ■

In the next example the elements of the time series are known only approximately and there is no obvious formula for the $n$th element of the sequence. We therefore turn to a graphical approach, suitable for use on a computer.

The idea is that if $f$ is an iterator mapping for a time series (5), then $f(x_n) = x_{n+1}$ and so

$$(x_n, x_{n+1}) = (x_n, f(x_n)).$$

This shows that the point $(x_n, x_{n+1})$ lies on the graph of $f$ and hence so do the points $(x_0, x_1), (x_1, x_2), (x_2, x_3), \ldots$ . Plotting a sufficiently large number of these points should therefore give a good approximation to the graph of $f$ (provided the points $x_0, x_1, x_2, x_3, \ldots$ are scattered liberally enough throughout the domain of $f$).

**2.2.3 Example** *The table below shows the first five elements of a sequence of numbers* $(x_0, x_1, x_2, x_3, \ldots)$, *generated by a computer. These numbers can be used, as in Figure 2.2.1, to plot the set of four points* $(x_0, x_1)$, $(x_1, x_2)$, $(x_2, x_3)$, $(x_3, x_4)$.

$$x_0 = 0.8, x_1 = 0.36, x_2 = 0.0784,$$
$$x_3 = 0.710986, x_4 = 0.178061.$$

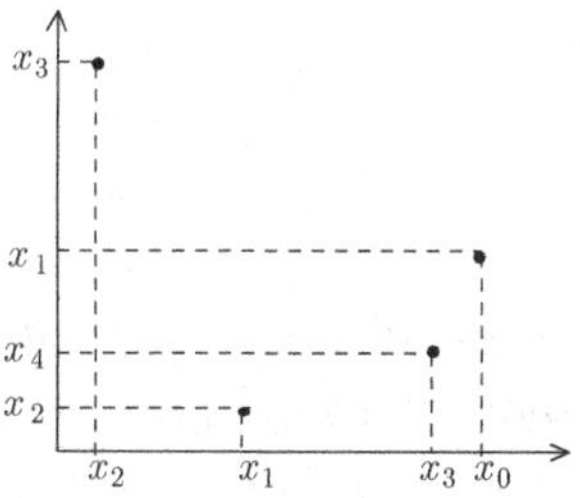

**Fig. 2.2.1** The sequence of numbers determines points in the plane. Do the points lie on the graph of a mapping $f$ ?

More elements of the sequence are stored in the computer, enabling it to plot the set of $n$ points

$$(x_0, x_1), (x_1, x_2), (x_2, x_3), \ldots, (x_{n-1}, x_n) \tag{7}$$

for many different values of $n$. Figure 2.2.2 shows some examples. As $n$ increases, the plots approximate the graph of a mapping — the iterator mapping of the original sequence.

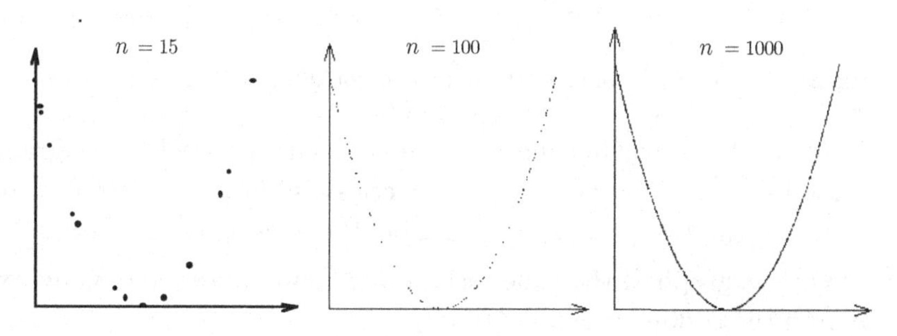


**Fig. 2.2.2** Computer plots of the points (7) for three values of $n$. ■

The sequence of numbers stored in the computer for use in the previous example was, in fact, just the sequence of iterates of 0.8 under the mapping $f$ where $f(x) = 4(x - 0.5)^2$. This explains the results in Figure 2.2.2. All we have achieved is to recover the original iterator mapping.

The intuitive idea of a sequence of random numbers is difficult to implement on a computer. Instead, computers produce *pseudo-random* sequences, which are almost as good in practice.

**2.2.4 Example** *In this example* $(x_0, x_1, x_2, \ldots)$ *denotes a specific pseudo-random sequence of numbers generated by the computer. Using these numbers the computer plotted the set of* $n$ *points*

$$(x_0, x_1), (x_1, x_2), (x_2, x_3), \ldots, (x_{n-1}, x_n) \tag{7}$$

*for three different values of* $n$. *The results are shown in Figure 2.2.3. Instead of forming part of the graph of a mapping, the plotted points are all over the unit square.*

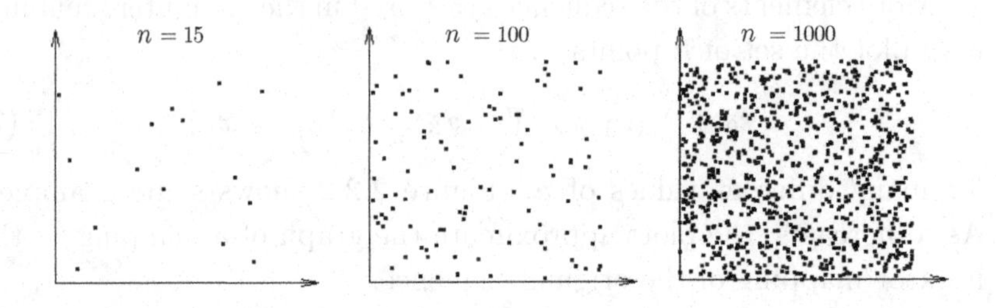


**Fig. 2.2.3** Computer plots of the points (7) for three different values of $n$. ■

In our discussion of time series, interest has centered on the graphs of the iterator mappings. The time series also have graphs. To avoid confusing the two graphs, we can label the axes as in Exercise 2.2.4:

On the graph of the time series $(x_n)_{n=1}^{\infty}$ we label the domain axis '$n$' and the codomain axis '$x_n$'.

On the graph of the iterator mapping $f$ we label the domain axis '$x_n$' and the codomain axis '$x_{n+1}$' (since $x_{n+1} = f(x_n)$).

## Exercises 2.2

2.2.1. Let $x_n$ be the $n$th element of the time-series $\left(1, \frac{1}{2}, \frac{1}{4}, \frac{1}{8}, \frac{1}{16}, \ldots\right)$.

(a) Plot the points $(x_0, x_1)$, $(x_1, x_2)$, $\ldots$, $(x_3, x_4)$.

(b) Does the time series have an iterator?

2.2.2. The time series in this exercise is $\left(1, -\frac{1}{2}, \frac{1}{4}, -\frac{1}{8}, \frac{1}{16}, - \ldots\right)$.

(a) Plot the graph of the $(n+1)$th element of the time series against the $n$th element (for $n = 0, 1, 2, 3$).

(b) Is there an iterator for this time series? If so, which mapping is it?

2.2.3. For each time series, decide whether there is an iterator:

(i) $\left(1, \frac{1}{2}, \frac{1}{3}, \frac{1}{3}, \frac{1}{2}, 1, \ldots\right)$ (ii) $\left(1, \frac{1}{2}, \frac{1}{3}, 1, \frac{1}{2}, \frac{1}{3}, \ldots\right)$.

2.2.4. The graphs below show (a) part of a time series and (b) a plot of its $(n+1)$th element against its $n$th element.

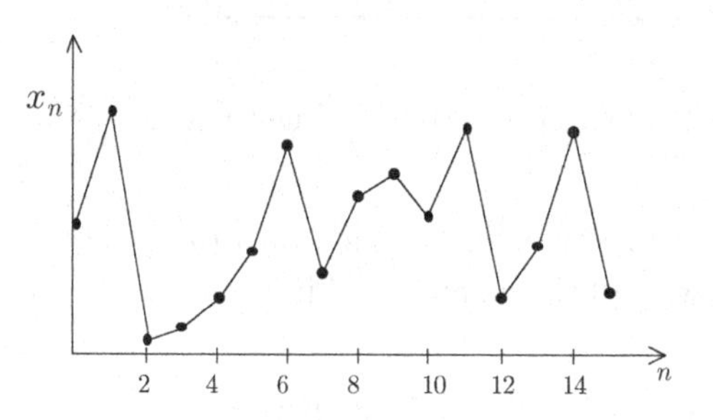


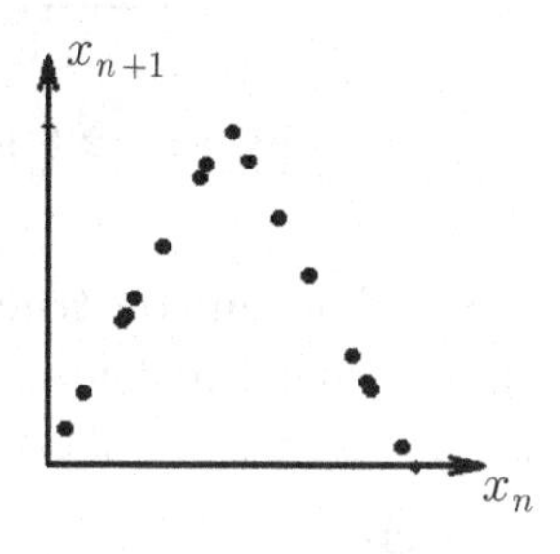


**Fig. 2.2.4** (a) Graph of time series. (b) Graph of iterator.

(a) Get rough estimates of the numbers $x_0, x_1, x_2, x_3$ from the graph on the left and then identify the points $(x_0, x_1)$, $(x_1, x_2)$ and $(x_2, x_3)$ in the plot on the right.

(b) Let $T$ denote the iterator mapping for the time series. Using the plot on the right, guess a formula for $T(x)$ in each case:

$$\text{(i)} \quad 0 \leq x \leq \tfrac{1}{2} \quad \text{and} \quad \text{(ii)} \quad \tfrac{1}{2} < x \leq 1.$$

## 2.3 ORBITS

Given a mapping $f : S \to S$ and a point $x_0 \in S$, iteration produces a sequence $x_0, x_1, x_2, \dots$ of elements of $S$. If $S \subseteq \mathbb{R}$ we can represent the sequence geometrically by sketching its graph. If $S$ consists of points in the plane, however, such graphs are not so relevant.

An alternative way to represent the iterates in such cases is to sketch the iterates as points in the set $S$. The order in which the iterates occur is then indicated by arrows showing where $f$ maps each point. This is illustrated in Figure 2.3.1.

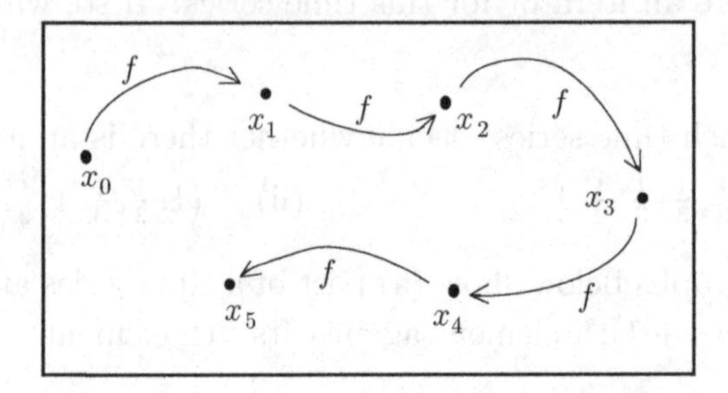


**Fig. 2.3.1** Portrait of the orbit of $x_0$ under $f$.

A diagram constructed in this way is called an *orbit portrait* . In this geometrical context, we call the sequence of iterates $(x_0, x_1, x_2, \dots)$ the *orbit* of $x_0$ under $f$.

Here we are thinking of an orbit as a sequence[2]

$$(x_0, x_1, x_2, \dots),$$

that is, as a *mapping* $n \mapsto x_n$ with the non-negative integers $\mathbb{N}_0 = \{0, 1, 2, \dots\}$ as domain. Thus for an orbit (as we define it) the order in which the elements occur is relevant and so are repetitions of any element.

[2]Our definition of orbit differs from that given by most authors. They define the orbit not as a sequence but as a *set* of elements

$$\{x_0, x_1, x_2, \dots\}$$

in which the order in which the elements occur is irrelevant.

**2.3.1 Example** *Find the orbit of the point* 1 *under the mapping* $f : \mathbb{C} \to \mathbb{C}$ *with* $f(z) = \frac{1}{2}iz$ *and sketch an orbit portrait.*

*Solution:* The orbit is $(1, \frac{1}{2}i, -\frac{1}{4}, -\frac{1}{8}i, \frac{1}{16}, \ldots)$, which spirals into the origin.

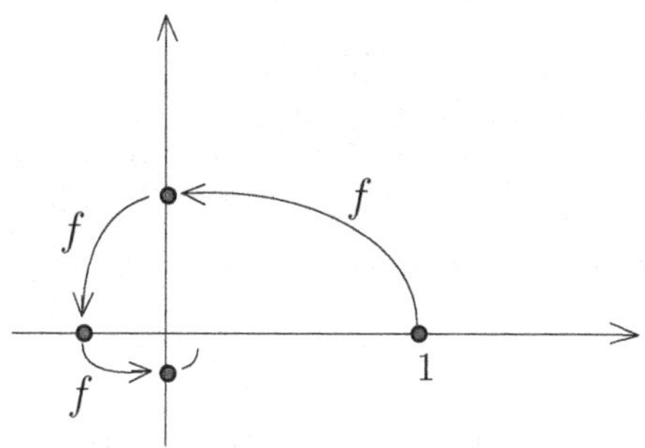


**Fig. 2.3.2** Portrait of an orbit in the complex plane $\mathbb{C}$.

**2.3.2 Example** *Let* $f : \mathbb{R} \to \mathbb{R}$ *with* $f(x) = 2x + 1$. *Find the orbit of the point* 1 *under* $f$ *and sketch both the graph of the sequence of iterates and the orbit portrait.*

*Solution:* The orbit is $(1, 3, 7, 15, \ldots)$, which goes off to infinity.

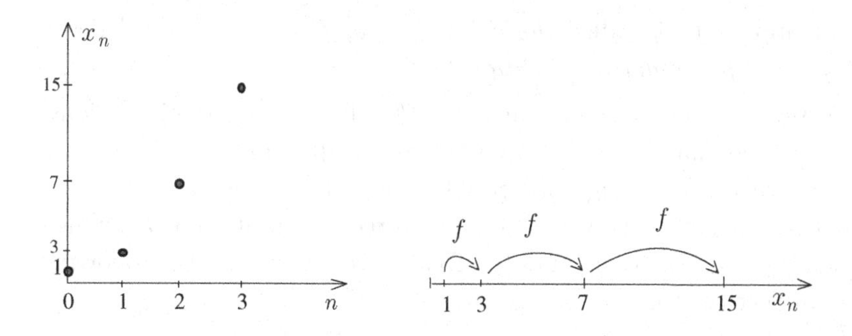


**Fig. 2.3.3** Graph of the orbit (left) and portrait of the orbit (right).

Orbit portraits bring out the role of the mapping $f$ in the iteration process. They show how the mapping $f$ is applied repeatedly, starting from the initial value, to produce the iterates.

The process of repeatedly applying the function $f$ can be expressed in terms of composition of functions. Since $f$ maps the set $S$ into itself so does each of the functions $f^2$, $f^3$, ... where

$$\begin{aligned} f^2 &= f \circ f, \\ f^3 &= f \circ f \circ f, \\ &\vdots \end{aligned}$$

In this notation, (3) becomes

$$\begin{aligned} x_1 &= f(x_0), \\ x_2 &= f(x_1) = f(f(x_0)) = f^2(x_0), \\ x_3 &= f(x_2) = f(f^2(x_0)) = f^3(x_0), \\ &\vdots \end{aligned} \tag{8}$$

In general we define the mapping $f^n : S \to S$ for each $n \in \mathbb{N}$ by putting

$$f^n = \underbrace{f \circ f \circ \cdots \circ f}_{n \text{ copies of } f}. \tag{9}$$

The mapping $f^n$ is called *the nth iterate of f*. It is also called the *nth power of f under composition.*[3]

Many of the claims we make about $f^n$ will seem intuitively obvious. On the other hand, even simple things can provide practice at the art of writing down rigorous proofs. In order to prove statements about iterates of maps, we need a more precise definition of $f^n$, which avoids the use of a string of dots. To achieve this we use the following consequence of (9):

$$\text{For every } n \in \mathbb{N}, \quad f^{n+1} = f \circ f^n. \tag{10}$$

[3]It should be noted that the notation $f^n$ is also used quite commonly for the $n$th power of $f$ under *multiplication* instead of composition. This is particularly so in elementary calculus where, for example, $\cos^2(x)$ denotes $\cos(x)\cos(x)$ rather than $\cos(\cos(x))$ . This is an ambiguity you must learn to live with.

We could make this into a *recursive definition* of $f^n$ by giving its definition when $n = 1$. We can do slightly better by starting at $n = 0$ and adopting the recursive definition

$$f^0 = \text{id} \; ;$$
$$\text{and for every} \quad n \in \mathbb{N}_0, \; f^{n+1} = f \circ f^n$$

where id is the identity mapping on the set $S$ (defined by $\text{id}(x) = x$ for all $x \in S$).

By using this definition it is possible to prove the following result, which is suggested by the table (8).

**2.3.3 Theorem** *For all $n \in \mathbb{N}_0$, the nth iterate of $x_0$ under $f$ is given by*

$$x_n = f^n(x_0).$$

**Proof.** Use mathematical induction in conjunction with the above recursive definition of $f^n$; see Exercise 2.3.4. ∎

The role of Theorem 2.3.3 is to reduce proofs of basic results about a sequence of iterates to results about $n$th powers of functions under composition. In this regard you should note that powers of $f$ under composition obey familiar exponent laws: for all $m, n \in \mathbb{N}$

$$f^n \circ f^m = f^{n+m} \qquad \text{and} \qquad (f^n)^m = f^{nm} \tag{11}$$

These laws should seem intuitively obvious. When each law is written out in full, there is on each side of the equality a number of copies of $f$ with a circle between each of them. Hence the first law says that $n$ copies put together with $m$ copies gives a total of $n + m$ copies. Similarly the second law asserts that $m$ blocks, each containing $n$ copies, contain a total of $mn$ copies.

**2.3.4 Example** *Find $f^2$ where $f : \mathbb{R} \to \mathbb{R}$ with $f(x) = 2x + 1$.*

*Solution:* $f^2(x) = f \circ f(x) = f(f(x)) = f(2x + 1) = 2(2x + 1) + 1$

Thus $f^2$ maps $\mathbb{R}$ to $\mathbb{R}$ with $f^2(x) = 4x + 3$. ∎

**2.3.5 Example** *Let $f : S \to S$ and, for each $n \in \mathbb{N}$, let $x_n$ be the nth iterate of $x_0 \in S$ under $f$. Show that for each $m \in \mathbb{N}$,*

$$x_{m+n} = f^m(x_n). \tag{12}$$

*Solution:* Let $m$ and $n$ be non-negative integers. By Theorem 2.3.3,

$$\begin{aligned} x_{m+n} &= f^{m+n}(x_0) \\ &= f^m \circ f^n(x_0) \\ &= f^m(f^n(x_0)) \\ &= f^m(x_n). \end{aligned}$$

■

We give another proof of (12) which makes more explicit use of the sequences of iterates involved. Let

$$(x_0, x_1, \ldots, x_{n-1}, x_n, x_{n+1}, \ldots, x_{m+n}, \ldots) \tag{13}$$

be the sequence of iterates of $x_0$ under $f$ and let

$$(x_n, x_{n+1}, \ldots, x_{n+m}, \ldots) \tag{14}$$

be the sequence (13) with its first $n$ elements struck out. In the sequence (14), just as in (13), we get from one element to the next by applying the mapping $f$. Hence $x_{m+n}$ is equal both to

(a) the result of applying $f$ to $x_0$ a total of $m + n$ times, and to

(b) the result of applying $f$ to $x_n$ a total of $m$ times.

The equality of (a) and (b) is just what the identity (12) asserts. ■

In terms of the difference equations in Chapter 1, the identity (12) says that: if (13) is a solution of a difference equation, then (14) is also.

In terms of population models, the identity (12) says that the increase in population depends only on the time which has elapsed, and not on the starting date. This is true provided environmental parameters such as the intrinsic reproduction rate $r$ and the carrying capacity $K$ remain constant over time.

## Exercises 2.3

2.3.1. Let $f : S \to S$.

(a) Verify that $f^2 \circ f^3 = f^5$ by writing each side out in full, without using the exponent notation.

(b) Verify that $(f^2)^3 = f^6$ in a similar way.

2.3.2. Let $f$ and $z_0$ be as in Example 2.1.4. Sketch an orbit portrait for $z_0$ under $f$.

2.3.3. Let $f : \mathbb{R} \to \mathbb{R}$ with $f(x) = 2x + 1$ (as in Example 2.1.2 ).

(a) Find a formula for $f^3(x)$, valid for each $x \in \mathbb{R}$.

(b) Guess a formula for $f^n(x)$ and prove your guess is valid for each $n \in \mathbb{N}$. Use mathematical induction.

2.3.4. Prove that for each $n \in \mathbb{N}$, the $n$th iterate of $x_0$ under $f$ is given by $x_n = f^n(x_0)$ (as claimed in Theorem 2.3.3). Use mathematical induction and the recursive definition (1) of $x_n$.

2.3.5. Let $f$ be a mapping. Assume that the sequence $(x_0, x_1, x_2, \dots)$ is a solution of the difference equation $x_{n+1} = f(x_n)$.

Show that, if $y_n = x_{2n}$, then the sequence $(y_0, y_1, y_2, \dots)$ is a solution of the difference equation $y_{n+1} = f^2(y_n)$.

2.3.6. Fronds of a plant depend on photosynthesis for their growth. Would you expect the increase in the number of fronds to depend only on the duration of the time which has the elapsed, but not on the starting time?

## Additional reading for Chapter 2

Composition of mappings and its relevance to iteration is explained in Section 4.2 of [PMJPSY] and (with less detail) in Section 1.3 of [De2]. Orbits are discussed in Section 1.4 of [De2].

The problem of reconstructing an iterator from its time series is introduced in Section 12.7 of [PJS].

To avoid the ambiguity inherent in the use of the notation $f^n$ for the $n$th iterate of a function $f$ under composition, [Ba] and [Gul] add some embellishments to this notation.

---

## References

[Ba] Michael F. Barnsley, *Fractals Everywhere,* 2nd edition, Academic Press Professional, Boston 1993.

[De2] Robert L. Devaney, *Chaos, Fractals and Dynamics, Computer Experiments in Mathematics,* Addison-Wesley, New York, 1990.

[Gul] Denny Gulick, *Encounters with Chaos,* McGraw-Hill, Inc., 1992.

[PJS] Hans-Otto Peitgen, Hartmut Jürgens and Dietmar Saupe, *Chaos and Fractals, New Frontiers of Science*, Springer-Verlag, New York, 1992.

[PMJPSY] Hans-Otto Peitgen, Evan Maletsky, Hartmut Jürgens, Terry Perciante, Dietmar Saupe, and Lee Yunker, *Fractals for the Classroom: Strategic Activities Volume Two,* Springer-Verlag, 1992. (two volumes)

# 3

# PERIODIC ORBITS

*Everyday life is governed by periodic processes. For example, we may catch the bus at the same time each morning or watch the news on TV at the same time every evening. In nature, species of migratory birds return to their breeding ground at the same time every year. In the physical world the sun rises and sets with reassuring regularity, while the tide ebbs and flows at regular intervals of time.*

*In Dynamical Systems the concept which corresponds to periodic processes is that of a* periodic orbit. *Recall that an orbit of a mapping is a sequence of elements from its domain. The orbit is said to be* periodic *if it consists of a block of $n$ elements repeated indefinitely. The positive integer $n$ is called a* period *of the orbit.*

*The simplest type of periodic orbit is one which has period one. Such an orbit consists of the same element repeated indefinitely.*

*The behaviour of a periodic orbit of period $n$ is known for all time once we know the first $n$ elements of the orbit. Thus the behaviour of a periodic orbit might seem to be regular and predictable. It is therefore surprising that periodic behaviour can occur alongside behaviour which is chaotic, as we show later.*

## 3.1 FIXED POINTS AND PERIODIC POINTS

**3.1.1 Definition** Let $S$ be a set and let $f : S \to S$ be a mapping. An element $x \in S$ is a *fixed point* of $f$ if $f(x) = x$. ■

If $x_0$ is a fixed point of $f$, then $f$ sends $x_0$ to itself. Successive applications of $f$ leave $x_0$ fixed, and so the portrait of the orbit of $x_0$ under $f$ is as in Figure 3.1.1.

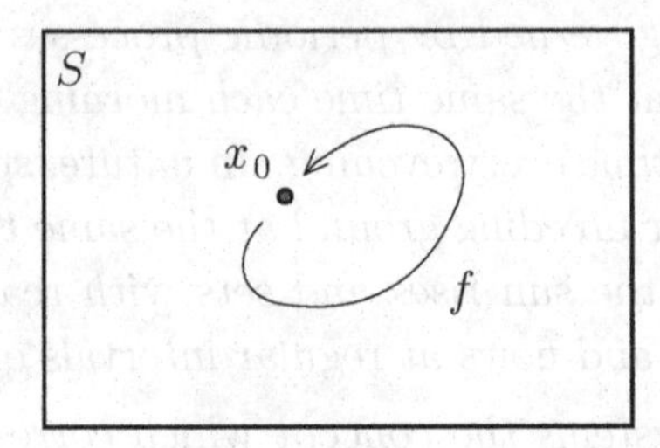


**Fig. 3.1.1** Orbit portrait for a fixed point.

Hence the $n$th iterate of $x_0$ under $f$ satisfies $x_n = x_0$ for all $n \in \mathbb{N}_0$ and so the sequence of iterates is the constant sequence $(x_0, x_0, x_0, \dots)$. The set of points in the orbit of $x_0$ contains the single point $x_0$.

In terms of difference equations: $x_0$ *is a fixed point for* $f$ if and only if $x_0$ *is the initial value for a constant solution of the difference equation* $x_{n+1} = f(x_n)$.

Fixed points of $f$ are obtained by solving the equation $x = f(x)$. If $f(x)$ is a simple enough expression, we can solve this equation algebraically.

**3.1.2 Example** *Find the fixed points of the quadratic mapping* $f : \mathbb{R} \to \mathbb{R}$ *with* $f(x) = x^2 - 2x - 4$.

*Solution:* The fixed points are the solutions of the equation $f(x) = x$; that is, of the equation $x^2 - 3x - 4 = 0$.

The solutions of this quadratic equation are $x = -1$ and $x = 4$. Thus the fixed points of the mapping $f$ are $-1$ and $4$. ■

We now generalize the idea of a fixed point.

**3.1.3 Definition** A point $x \in S$ is a *periodic point* of $f : S \to S$ if $f^n(x) = x$ for some $n \in \mathbb{N}$.

When this happens, the integer $n$ is called *a period* of the point $x$. The smallest such positive integer is called *the prime period* of $x$. A point of period $n$ is called *a period-$n$ point.* ■

Here are some simple consequences of the definition:

A period-$n$ point of $f$ is a fixed point of $f^n$.
In particular, a period-1 point of $f$ is a fixed point of $f$.
A period-1 point is a period-$n$ point for every $n \in \mathbb{N}$.

The following example illustrates the definition in the case $n = 2$.

**3.1.4 Example** *Let $f$ be as in Figure 3.1.2. Why is $x_0$ a periodic point of $f$? What is the prime period of $x_0$? What is the orbit of $x_0$ under $f$? What is the orbit of $x_1$?*

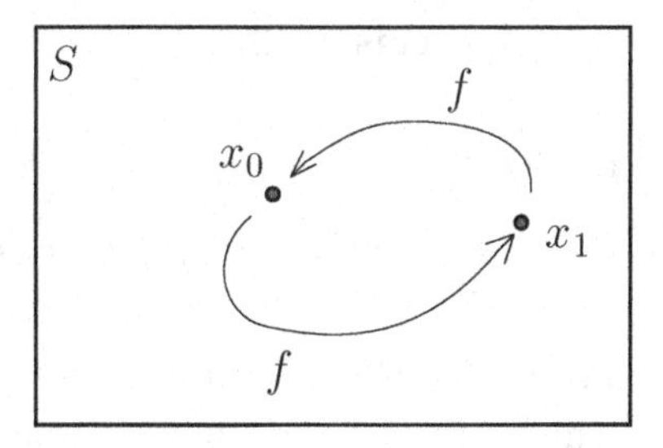


**Fig. 3.1.2.** Orbit portrait for a period-2 point.

*Solution:*

Since $f^2(x_0) = x_0$ the point $x_0$ is periodic and 2 is a period. Since also $f(x_0) \neq x_0$, the smallest period of $x_0$ is 2. Hence the prime period of $x_0$ is 2.

The orbit of $x_0$ is the sequence $(x_0, x_1, x_0, x_1, x_0, x_1, \ldots)$ and the orbit of $x_1$ is the sequence $(x_1, x_0, x_1, x_0, x_1, x_0, \ldots)$. ■

Notice that in the above example the one orbit portrait gives rise to two different orbits. One of the orbits has $x_0$ as the initial value while the other orbit starts with $x_1$.

If $x_0$ has period $n$ then

$$\begin{aligned} x_n &= f^n(x_0) = x_0, \\ x_{n+1} &= f^{n+1}(x_0) = f(f^n(x_0)) = f(x_0) = x_1, \\ x_{n+2} &= f^{n+2}(x_0) = f(f^{n+1}(x_0)) = f(x_1) = x_2, \\ &\vdots \end{aligned}$$

Thus, after $n$ iterations, the points in the orbit start to repeat themselves. Hence the sequence of iterates consists of the first $n$ elements repeated indefinitely:

$$(x_0, x_1, x_2, \ldots, x_{n-1}, x_0, x_1, x_2, \ldots, x_{n-1}, \ldots).$$

Each of the points in the orbit, moreover, is itself a point of period $n$.

## Exercises 3.1

3.1.1. Find the fixed points of $f : \mathbb{R} \to \mathbb{R}$ in each case:

(a) $f(x) = x^2$ (b) $f(x) = -x$ (c) $f(x) = \begin{cases} 1/x & \text{if } x \neq 0 \\ 0 & \text{if } x = 0. \end{cases}$

3.1.2. Find the points of period 2 of the function $f : \mathbb{R} \to \mathbb{R}$ in each case and state which of the points have prime period 2:

(a) $f(x) = x^3$ (b) $f(x) = -x^3$.

3.1.3. Let $S$ be a set. Give an example of a map $f : S \to S$ for which all of the points in S are fixed points. How many such mappings are there?

3.1.4. Give an example of a set $S \subseteq \mathbb{R}$ and a mapping $f : S \to \mathbb{R}$ for which all the points of $S$ have prime period 2.

3.1.5. Let $x_0$ be a periodic point of a mapping, with prime period 2.

(a) Is $x_0$ a period-3 point of the mapping?

(b) Is $x_0$ a period-4 point of the mapping?

(c) Give reasons for your answers.

3.1.6. Let $f : \mathbb{R} \to \mathbb{R}$ with $f(x) = x + 1$. Show that $f$ has no fixed points. Show, furthermore, that $f$ has no periodic points of any period $n \in \mathbb{N}$.

3.1.7. Let $x_0, x_1, x_2$ be three distinct points in the domain of a mapping $f$. Suppose, furthermore, that $f$ maps the set of these three points into itself and that $f$ has no points with period-2. Sketch the possible orbit diagrams for $x_0$ under $f$.

3.1.8. Repeat Exercise 7, but this time suppose that $f$ can have points of period-2, but has none of prime period 3.

3.1.9. Let $x$ be a fixed point of $f : S \to S$.

(a) Show that $x$ is a fixed point of $f^2$.

(b) Prove that $x$ is a fixed point of $f^n$ for each $n \in \mathbb{N}$. Use mathematical induction.

3.1.10. Let $x$ be a periodic point of a mapping $f : S \to S$. Suppose that $n$ is a period and $p$ is the prime period of $x$. Verify that $n$ is an integer multiple of $p$.

[Hint: By the division algorithm (see [Fr], page 53) there exist integers $q \geq 0$ and $r \geq 0$ such that $n = pq + r$ and $0 \leq r < p$. Prove that $r = 0$.]

3.1.11. Express the result of Exercise 10 in words without using any mathematical symbols.

3.1.12. Let $x_0$ be a periodic point of period $n$. Prove that each of the points in the orbit of $x_0$ also has period $n$.

[Hint. Apply the result of Exercise 9(b) to $f^n$.]

3.1.13. Let $x_0$ be a periodic point of period $n$.

(a) Prove that for each $m \in \mathbb{N}$ there is an $r \in \mathbb{Z}$ such that

$$x_m = x_r \quad \text{and} \quad 0 \leq r < n.$$

(b) What does (a) tell you about the number of distinct points in the orbit of $x_0$?

[Hint for (a). By the division algorithm, there are integers $s \geq 0$ and $r \geq 0$ such that $m = ns + r$ and $0 \leq r < n$. ]

## 3.2 FINDING FIXED POINTS

Finding the fixed points of a mapping $f : S \to S$ involves solving the equation

$$f(x) = x \qquad (1)$$

If $f(x)$ is simple enough, then (1) can be solved using elementary algebra. For more complicated equations, however, we resort to the use of graphs.

### Using algebra

If $f(x)$ is quadratic in $x$, then the equation (1) is a quadratic equation and hence can be solved by the well-known quadratic formula.

Although there are analogous formulae for solving polynomial equations of degree 3 or 4, which involve taking cube roots and fourth roots, the formulae are not easy to apply in practice. For polynomial equations whose degree exceeds 4, moreover, there are no such formulae (as was proved in the nineteenth century by the famous French mathematician Évariste Galois.)

There are, nevertheless, special cases of polynomial equations of degree higher than two, for which we can obtain the solutions with the aid of elementary algebra. In such cases, the following theorem[1] is often useful. (The theorem uses the following nomenclature: a number $a$ is a *zero* of a polynomial $p(x)$ if $p(a) = 0$.)

**Factor Theorem.** If $p(x)$ is a polynomial of degree $n \geq 1$ and has the numbers $a_1, a_2, \ldots, a_k$ (all distinct) as zeroes, then

$$p(x) = (x - a_1)(x - a_2)\ldots(x - a_k)q(x)$$

where $q(x)$ is a polynomial of degree $n - k$.

[1]This very useful theorem remains nameless in the algebra books. It is usually found in the vicinity of things like 'the Division Algorithm' or the 'Remainder Theorem'.

**3.2.1 Example** *Let $f : \mathbb{R} \to \mathbb{R}$ with $f(x) = x^2 - 2$.*

*(a) Find the period-1 points of $f$.*

*(b) Find the period-2 points of $f$.*

*(c) Give two different orbits of prime period two.*

*Solution:*

(a) The period-1 points are the solutions of the equation $f(x) = x$; that is, of $x^2 - x - 2 = 0$. The solutions of this quadratic equation are $x = -1$ and $x = 2$. Thus the period-1 points are $-1$ and 2.

(b) The two period-1 points just found are also period-2 points. We can use this fact to help find the remaining period-2 points. The period-2 points of $f$ are the fixed points of $f^2$; that is they are solutions of the fourth degree equation $f^2(x) = x$, which is

$$\begin{aligned} (x^2-2)^2 - 2 &= x, \\ x^4 - 4x^2 + 2 &= x, \\ x^4 - 4x^2 - x + 2 &= 0. \end{aligned} \tag{2}$$

But $x = -1$ and $x = 2$ are solutions of (2) as $-1$ and 2 are period-2 points. By the Factor Theorem, the polynomial on the LHS of (2) factorizes to give

$$\begin{aligned} x^4 - 4x^2 - x + 2 &= (x+1)(x-2)q(x) \\ &= (x^2 - x - 2)q(x) \end{aligned}$$

where $q(x)$ is a quadratic. We can now find $q(x)$ by equating coefficients (see Exercise 3.2.1) or by long division. Either way we get $q(x) = x^2 + x - 1$. The remaining solutions of (2) are the solutions of $q(x) = 0$. These are $x = a_1$ and $x = a_2$ where

$$a_1 = \frac{-1-\sqrt{5}}{2} \quad \text{and} \quad a_2 = \frac{-1+\sqrt{5}}{2}.$$

Thus the period-2 points are $-1$, 2, $a_1$ and $a_2$.

(c) One prime period-2 orbit is $(a_1, a_2, a_1, a_2, \ldots)$.
Another prime period-2 orbit is $(a_2, a_1, a_2, a_1, \ldots)$. ∎

## Using graphs

Recall that a fixed point $x$ of a mapping $f : S \to S$ is a solution of the equation

$$f(x) = x.$$

The number of mappings for which one can find the periodic points by using elementary algebra is very small, and so we turn to graphical methods, which are more generally applicable when $S \subseteq \mathbb{R}$. We first consider the practical aspects, and later consider how to give theoretical justification for what we do.

We use the notation $\mathrm{id}\colon S \to S$ with $\mathrm{id}(x) = x$ to denote the identity mapping on the set $S$. The equation defining fixed points can then be written as

$$f(x) = \mathrm{id}(x). \tag{3}$$

From equation (3) it follows that *$x$ is a fixed point of $f$* if and only if

> *the graphs of $f$ and* id *have a point of intersection on the vertical line through the point $x$ on the domain axis.*

The relevant points are shown in Figure 3.2.1.

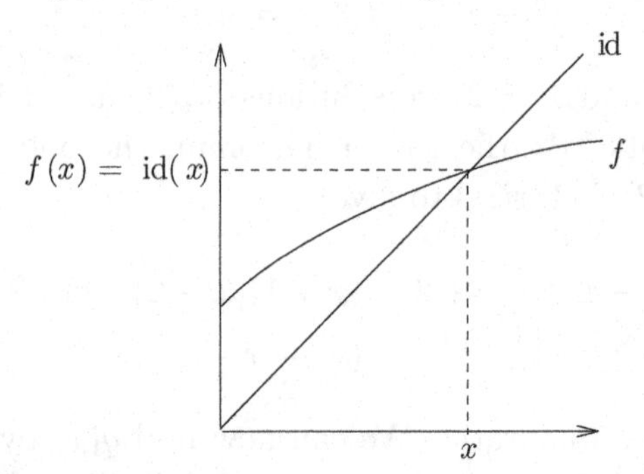


**Fig. 3.2.1** The graphs of $f$ and id intersect at a point on the vertical line through the point $x$ on the domain axis. Hence $f(x) = \mathrm{id}(x) = x$ and so $x$ is a fixed point of $f$.

We illustrate the graphical method by applying it to the same mapping as in Example 3.2.1, where we found the fixed points algebraically. Note that since period-1 points are also period-2 points, the intersection points of the graphs of $f^2$ and id include those of $f$ and id.

**3.2.2 Example** *Let $f : \mathbb{R} \to \mathbb{R}$ with $f(x) = x^2 - 2$. Use graphs to find approximately*

(a) *the period-1 points of $f$*

(b) *the period-2 points of $f$.*

*Solution:*

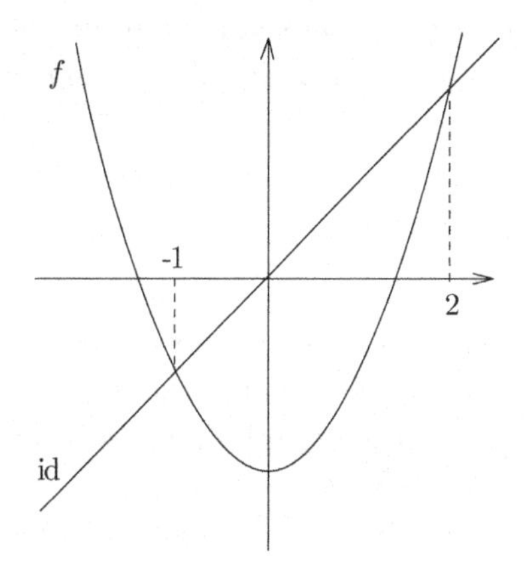


**Fig. 3.2.2(a)**

The graphs of $f$ and id intersect in two points with $x$-coordinates approximately

$$-1.0 \quad \text{and} \quad 2.0$$

respectively. These two numbers are the period-1 points of $f$ (which were found algbraically in Example 3.2.1).

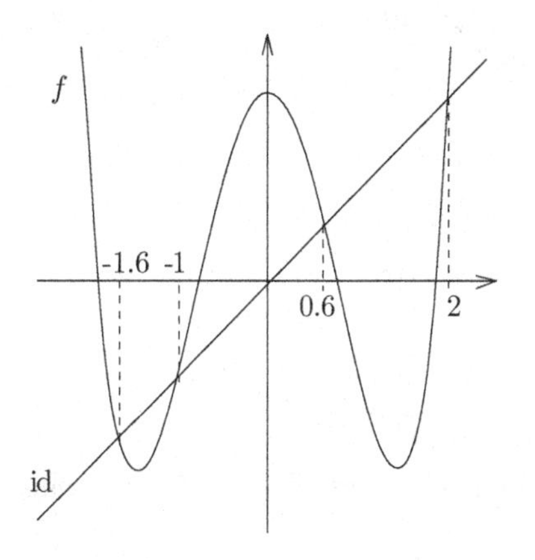


**Fig.3.2.2(b)**

The graphs of $f^2$ and id intersect in four points.

The $x$-coordinates of two of these points are approximately

$$-1.0 \quad \text{and} \quad 2.0.$$

respectively. These two points are the period-1 points found above.

The other two of these points have $x$-coordinates approximately

$$-1.6 \quad \text{and} \quad 0.6.$$

respectively. These are the points $a_1$ and $a_2$ with prime period 2 (found by algebra in Example 3.2.1).

By zooming in on small portions of the graph and sketching them more accurately, on a larger scale, we can find the periodic points as accurately as we want. ∎

### Justifying the use of graphs

A computer sketches a graph using approximations to the values of the function. A computer can tell us whether the value of a function at a given point is approximately zero, but it cannot tell us that the value is exactly zero. Hence we can never be sure that any given point is the *exact* point of intersection of the graphs of $f$ and id.

> *How then can we be certain from the graphs that there are indeed exact fixed points and not merely ones which are approximately fixed ?*

The answer to this question is based on the fact that if the value of a function is *nonzero* at some point, then this can be confirmed by a computer working to sufficient accuracy.

Suppose for example that $f$ is a continuous function on a domain which includes an interval $[a, b]$ and that at the point $a$ the graph of $f$ lies above that of id, while at the point $b$ the graph of $f$ lies below that of id (as in Figure 3.2.3).

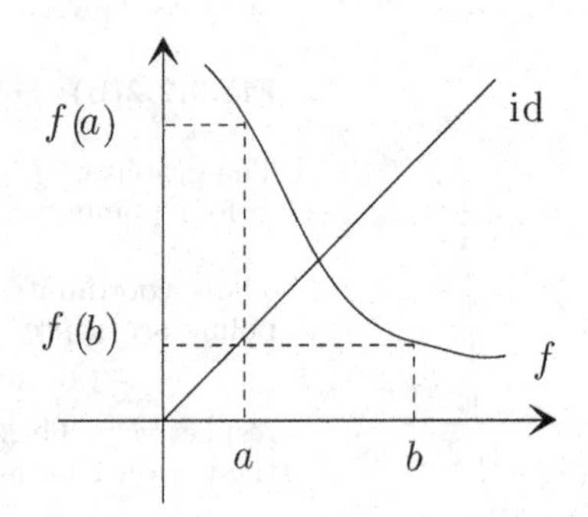


**Fig. 3.2.3** Graph of $f$ crossing that of id.

If the computer is working to sufficient accuracy we can be sure that

$$f(a) - \mathrm{id}(a) > 0 \qquad \text{while} \qquad f(b) - \mathrm{id}(b) < 0.$$

Thus if we put

$$\phi = f - \mathrm{id}$$

then the values $\phi(a)$ and $\phi(b)$ have opposite signs.

Since $f$ is a continuous function, then so is $\phi$, and hence we may apply the following theorem from calculus.

**Intermediate Value Theorem.** If $\phi$ is a real-valued function which is continuous at each point of a closed interval $[a, b]$, then for each $y$ between $\phi(a)$ and $\phi(b)$ there is an $x \in [a, b]$ such that $\phi(x) = y$.

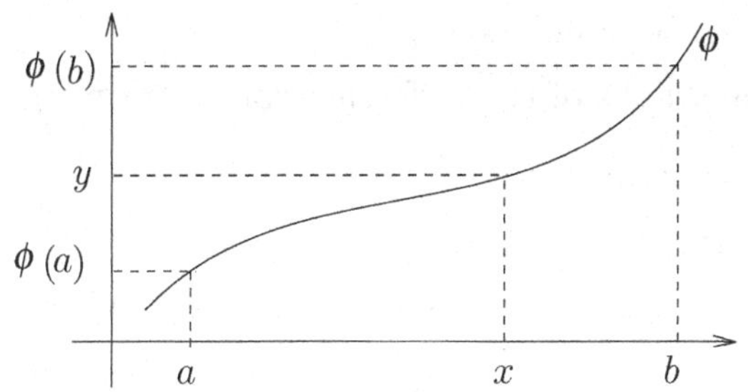


**Fig. 3.2.4** If $y$ is between $\phi(a)$ and $\phi(b)$, then there is an $x$ between $a$ and $b$ which $\phi$ maps to $y$.

In our problem, the numbers $\phi(a)$ and $\phi(b)$ have opposite signs and so 0 lies between them. Hence we may choose $y = 0$ in the Intermediate Value Theorem. We may therefore conclude that there is a point $x$ between $a$ and $b$ such that $\phi(x) = 0$; that is, $f(x) = x$.

Thus we have established the existence of a number $x$ which is a fixed point of $f$.

## Exercises 3.2

3.2.1. Find a quadratic $q(x)$ such that, for all $x \in \mathbb{R}$,

$$x^4 - 4x^2 - x + 2 = (x^2 - x - 2)q(x).$$

[Hint. Write $q(x) = ax^2 + bx + c$, expand the right-hand side and then equate coefficients of powers of $x$ to get simultaneous equations for $a, b$ and $c$.]

3.2.2. Is there a quadratic $q(x)$ such that, for all $x \in \mathbb{R}$,

$$x^4 - 4x^2 - x + 2 = (x^2 + x - 2)q(x) \ ?$$

3.2.3. Let $f : \mathbb{R} \to \mathbb{R}$ and let $f^2$ intersect the graph of id in just one point. How many points of period 1 does $f$ have?
In how many points does the graph of $f$ intersect that of id?

3.2.4. Let $f : \mathbb{R} \to \mathbb{R}$ with $f(x) = x^2 - 6$.
(a) Find the period-1 points of $f$.
(b) Find the period-2 points of $f$.
(c) Sketch the graphs of two different orbits of prime period-2.

3.2.5. Let $f(x)$ be quadratic in $x$.
(a) What are the degrees of the polynomials $f^2(x), f^3(x), f^4(x)$?
(b) What is the degree of the polynomial $f^n(x)$, where $n \in \mathbb{N}$?

3.2.6. Let $f(x)$ be a polynomial of degree $d \geq 1$ and suppose that $f$ has $k$ distinct fixed points $a_1, a_2, \ldots, a_k$. Prove that for each $n \in \mathbb{N}$,

$$f^n(x) = x + (x - a_1)(x - a_2)\ldots(x - a_k)g(x)$$

where $g(x)$ is a polynomial. What is the degree of $g(x)$?

3.2.7. Sketch the graphs of the functions cos and id on the one diagram and then use the Intermediate Value Theorem to prove that cos has a fixed point between 0 and $\pi/2$.

3.2.8. Copy the graph shown in Figure 3.2.5 and then show a point $y$ between $\phi(a)$ and $\phi(b)$ such that the equality $y = \phi(x)$ holds for just one number $x$ between $a$ and $b$.

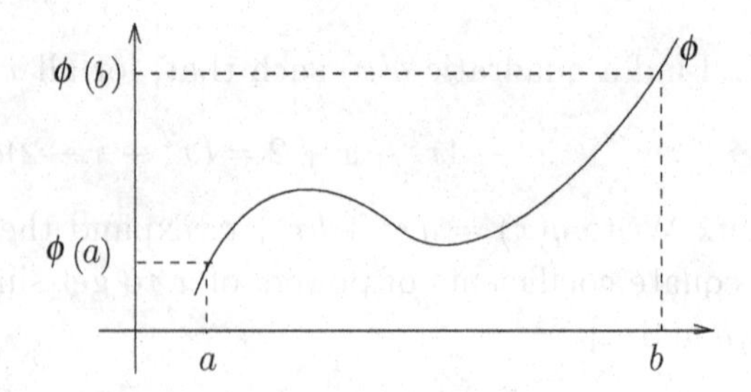


**Fig. 3.2.5.**

3.2.9. Repeat Exercise 8 but with the requirement that $y = \phi(x)$ holds for exactly three numbers $x$ between $a$ and $b$.

## 3.3 EVENTUALLY PERIODIC ORBITS

**3.3.1 Definition** Let $f : S \to S$. A point $x_0 \in S$ is said to be an *eventually periodic point* of $f$ if some iterate $x_k$ of $x_0$ is a periodic point of $f$.

A point is said to have *eventual period* $n$ if some iterate of the point has period $n$. ∎

Every periodic point is eventually periodic (in the definition we may take $k = 0$). A point is said to be eventually fixed if it has eventual period 1.

**3.3.2 Example** *Show that $1 + \sqrt{3}$ is an eventually periodic point of the mapping $f : \mathbb{R} \to \mathbb{R}$ with $f(x) = x^2 - 2x - 4$. Sketch an orbit portrait and also the graph of the orbit.*

*Solution:* A little arithmetic gives

$$\begin{aligned} f(1 + \sqrt{3}) &= -2, \\ f(-2) &= 4, \\ f(4) &= 4. \end{aligned}$$

Thus the second iterate of $1 + \sqrt{3}$ is 4 and this is a periodic point. Hence $1 + \sqrt{3}$ is an eventually periodic point.
The orbit of $1 + \sqrt{3}$ is $(1 + \sqrt{3},\ -2,\ 4,\ 4,\ 4, \ldots)$.

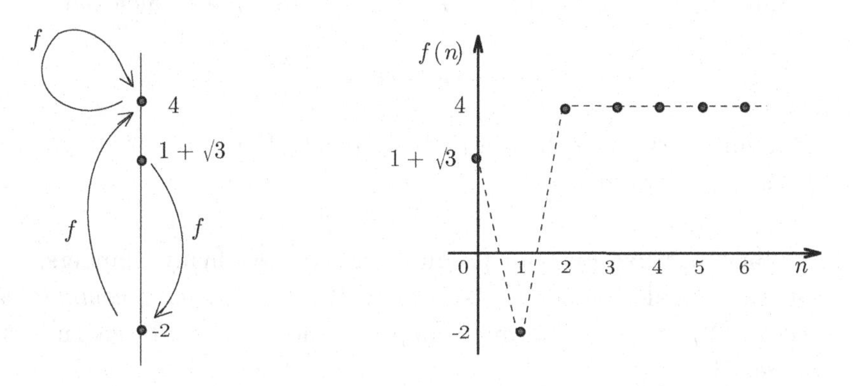


**Fig. 3.3.1** Portrait of the orbit (left) and graph of the orbit (right). ∎

**3.3.3 Example** *Let $f$ be as in Figure 3.3.2. What is the orbit of $x_0$? What is the orbit of $x_2$? Why is $x_0$ an eventually periodic point? What is an eventual period of $x_0$?*

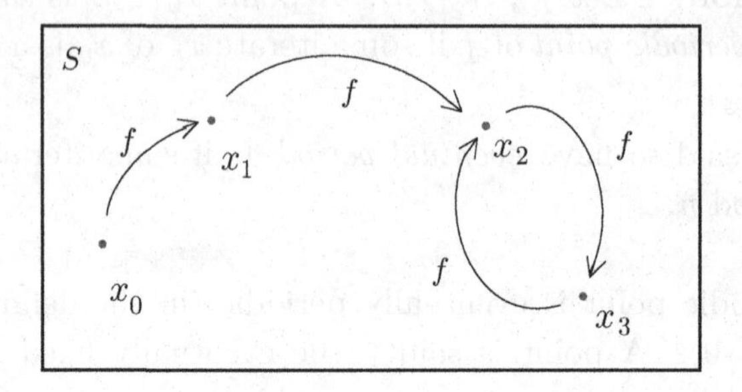


**Fig. 3.3.2** An eventually periodic orbit .

*Solution:*
The orbit of $x_0$ is $(x_0, x_1, x_2, x_3, x_2, x_3, \ldots)$.
The orbit of $x_2$ is $(x_2, x_3, x_2, x_3, x_2, x_3, \ldots)$.
The point $x_0$ is eventually periodic because one of its iterates, $x_2$, is a periodic point. The number 2 is an eventual period of $x_0$. ∎

**3.3.4 Example** *Suppose that $f : S \to S$ where $S$ is a set with only finitely many elements. Prove that every $x_0 \in S$ is an eventually periodic point.*

*Solution:* Let the orbit of $x_0$ under $f$ be $(x_0, x_1, x_2, x_3, \ldots)$.
Since $S$ has only finitely many points, not all the points in this orbit can be distinct. Thus there is a $k \in \mathbb{N}_0$ and an $n \in \mathbb{N}$ such that

$$x_k = x_{n+k}.$$

But this shows that $x_k$ is a periodic point. Hence the orbit of $x_0$ is eventually periodic. ∎

Since a computer can produce only finitely many numbers, the above example implies that *every orbit generated by a computer is eventually periodic!* The eventual period, however, can be enormously large.

Recall that a mapping is said to be *one-to-one* if it never maps two different points in its domain to the same value in its codomain. In Exercise 3.3.3 you will show that for such a mapping there are no eventually periodic points other than the periodic points themselves.

## Exercises 3.3

3.3.1. Let $f : \mathbb{R} \to \mathbb{R}$ be given by $f(x) = 4x(1-x)$.

(a) Prove that each of the following points is an eventually fixed point of $f$: $1, \quad 1/2, \quad 1/2 + 1/(2\sqrt{2})$

(b) Sketch an orbit portrait and a graph for each of the eventually fixed points found in (a).

(c) Find another eventually fixed point of $f$.

3.3.2.

(a) In Example 3.3.4, what is the *hypothesis* ? what is the *conclusion* ? Hence write the result as an *if ... then* statement.

(b) Write down the *converse* of the *if ... then* statement from (a).

(c) Is the statement you have written in (b) true or false? Give reasons for your answer.

3.3.3. Let $f : S \to S$ be a one-to-one mapping and let $x_0$ be an eventually periodic point of $f$. Let $(x_0, x_1, x_2, \ldots)$ be the orbit of $x_0$ under $f$. You will show below that this orbit is periodic.

(a) State why there is an element $k \in \mathbb{N}_0$ such that $x_k$ is a periodic point of $f$.

Let $k$ be the smallest such element of $\mathbb{N}_0$.
Let $n \in \mathbb{N}$ be a period of $x_k$.
Assume $k > 0$. Hence $x_{k-1}$ is defined.

(b) Why is $x_{k-1} \neq x_{k+n-1}$?

(c) State why $f(x_{k-1}) = x_k$ and $f(x_{k+n-1}) = x_k$.
Deduce that $x_{k-1} = x_{k-1+n}$.

(d) Thus the assumption $k > 0$ leads to a contradiction.
What follows about $k$? about the orbit of $x_0$?

## Additional reading for Chapter 3

Periodic and eventually periodic points are discussed in Section 1.5 of [De2] and in Sections 1.2 and 1.3 of [Gul].

A precise statement of what Galois proved about polynomial equations may be seen in [Fr] or [Sti].

A proof of the Intermediate Value Theorem can be found in any good text book on calculus. See for example [Ap].

## References

[Ap] Tom Apostol, *Calculus,* Blaisdell Publishing Company (1961).

[De2] Robert L. Devaney, *Chaos, Fractals and Dynamics, Computer Experiments in Mathematics,* Addison-Wesley, New York, 1990.

[Fr] John B. Fraleigh, *Introduction to Abstract Algebra (2nd edition)*, Addison-Wesley, Reading, Massachusetts, 1978.

[Gul] Denny Gulick, *Encounters with Chaos,* McGraw-Hill, Inc., 1992.

[Sti] John Stillwell, "Galois Theory for Beginners", *American Mathematical Monthly,* **101** (1994), 22-27.

# 4

# ASYMPTOTIC ORBITS I
## Linear and affine mappings

*In the previous chapter graphs were used to study the periodic orbits of a mapping* $f : S \to S$ *in the special case that* $S$ *is a subinterval of* $\mathbb{R}$. *In particular we showed how the points of intersection of the graphs of* $f$ *and* id *give the period-1 points of* $f$. *In this chapter we use these graphs to study more complicated orbits.*

*We show how to represent the successive iterates of a point on a graph by introducing the idea of a cobweb diagram. Cobweb diagrams will be studied first for linear and affine mappings. This leads us to study orbits which converge to, or which diverge away from, a fixed point. Such orbits are said to be asymptotic to the fixed point.*

*We prove theorems on the dynamics of linear and affine mappings. Later we show how the dynamics of affine mappings can be used to approximate the dynamics of differentiable mappings near a fixed point.*

## 4.1 COBWEB DIAGRAMS

Let $f : S \to S$ where $S$ is a subinterval of the real numbers and let $x_0 \in S$. Given the graph of $f$, how can we use it to produce the sequence of iterates $x_0, x_1, x_2, x_3, \ldots$ of $x_0$ under $f$ ?

Well, the first iterate is easy: the usual construction of a typical point $(x_0, f(x_0))$ on the graph of $f$ is illustrated in Figure 4.1.1. This puts the first iterate $x_1 = f(x_0)$ on the codomain axis.

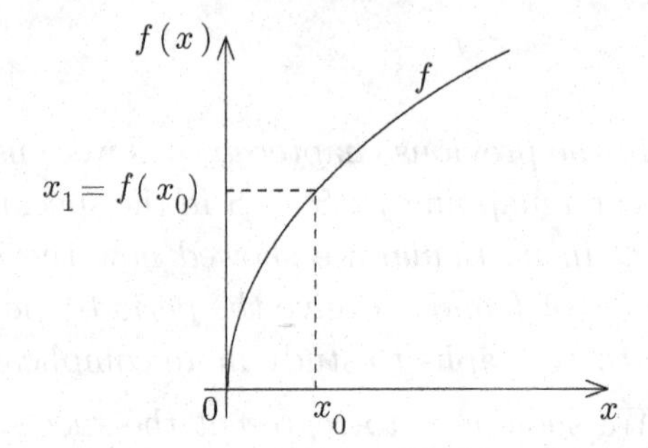


**Fig. 4.1.1**
The first iterate of $x_0$ can be read directly from the graph.

Before constructing the second iterate, however, it is necessary to bring $x_1$ back onto the domain axis. This could be achieved by reflection in the diagonal (that is, in the graph of id) as shown in Figure 4.1.2. Once $x_1$ is back on the domain axis, we can apply $f$ once again to get $x_2 = f(x_1)$, and so on.

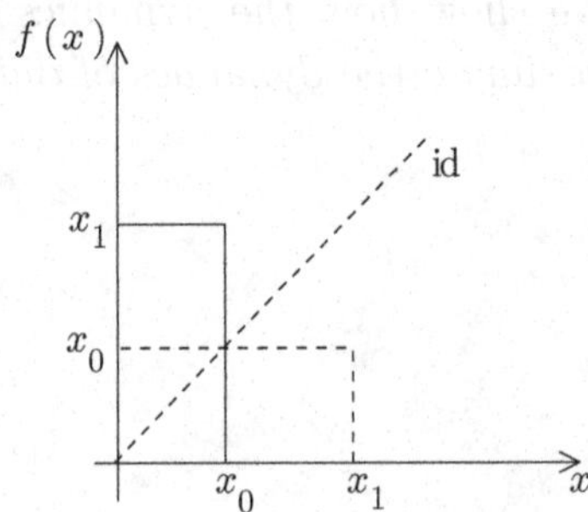


**Fig. 4.1.2**

Bringing the first iterate back onto the domain axis.

This procedure has the disadvantage, however, of making us forever chop and change between the two axes. A more systematic procedure is to represent the successive iterates by points on the *diagonal*. This keeps all the iterates on the one line in a very natural way.

Thus, instead of using $x_0$ to produce $x_1$, we work on the diagonal and use $(x_0, x_0)$ to produce $(x_1, x_1)$. The procedure for doing this, which is illustrated in Figure 4.1.3, is as follows:

1. *Plot the point* $(x_0, x_0)$ *on the graph of* id.
2. *From this point on the graph of* id,
   *move in a vertical direction until you reach the graph of* $f$,
   *at the point* $(x_0, f(x_0)) = (x_0, x_1)$.
3. *From this point on the graph of* $f$,
   *move in a horizontal direction until you reach the graph of* id,
   *at the point* $(x_1, x_1)$.

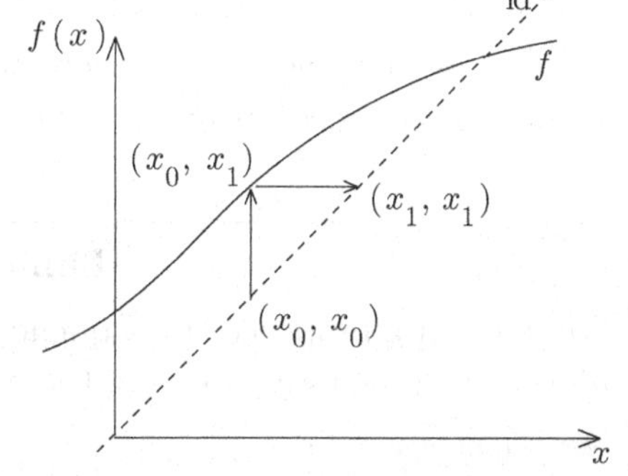


**Fig. 4.1.3**
One step of the iteration.

The same procedure which produces $(x_1, x_1)$ from $(x_0, x_0)$ can be used to produce $(x_2, x_2)$ from $(x_1, x_1)$, and so on. Continuing in this way indefinitely gives the sequence $(x_0, x_0), (x_1, x_1), (x_2, x_2), \ldots$ of points on the graph of id (see Figure 4.1.4). Projecting these points onto the domain axis then gives the iterates $x_0, x_1, x_2, \ldots$

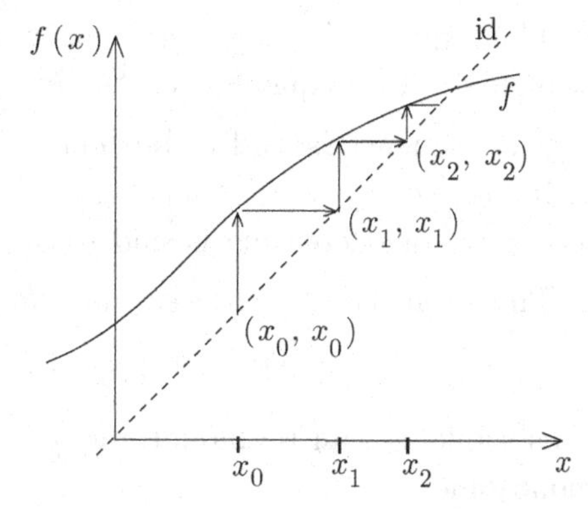


**Fig 4.1.4**
Repetition of the step.

## Dynamics

In studying the dynamics of a mapping we are interested in such questions as the following.

*What is the long-term behaviour of the orbits of the mapping?*
*What is their ultimate fate?*
*Do they keep close to where they started or do they move away?*
*Does an orbit get close to a periodic orbit?*
*Is it eventually periodic?*

Cobweb diagrams are a useful tool for suggesting answers to these types of questions since they enable us to produce sequences of iterates geometrically and to display them on one of the axes.

To express our answers precisely, we shall use the language of convergence of sequences of real numbers from calculus.

**Limits**

We say that a property of a sequence $(x_0, x_1, x_2, \dots)$ holds *eventually* if, for some integer $n \geq 0$, the property holds for the sequence $(x_n, x_{n+1}, x_{n+2}, \dots)$.

The sequence is said to *converge to a limit* $\ell \in \mathbb{R}$ if, for every number $\epsilon > 0$, the elements of the sequence are all eventually within a distance $\epsilon$ of the limit $\ell$. We then write

$$\lim_{n \to \infty} x_n = \ell.$$

The sequence is said to *diverge to* $\infty$ if for each real number $b$ the elements of the sequence eventually all exceed $b$.

The sequence is said to be *increasing* if $x_0 \leq x_1 \leq x_2 \leq \dots$ and to be *decreasing* if $x_0 \geq x_1 \geq x_2 \geq \dots$ . A sequence which is either increasing or decreasing is said to be *monotonic*.

The *sequence of distances from the limit* is

$$(|x_0 - \ell|, |x_1 - \ell|, |x_2 - \ell|, \dots).$$

This sequence can be monotonic, even if the original sequence is not monotonic.

## Exercises 4.1

4.1.1. The two diagrams in Figure 4.1.5 both purport to show some iterates of a point under $f$. Which is correct? Why?

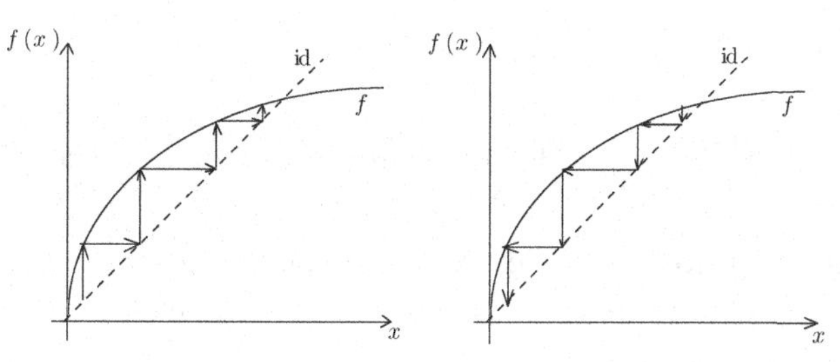


**Fig. 4.1.5** Which of these diagrams is correct?

4.1.2. Sketch the graphs of id and $f : [0, \infty) \to [0, \infty)$ with $f(x) = x^2$. What are the fixed points of $f$? In each of the following cases use a cobweb diagram to help you guess the given set of initial values.

(a) $\{x_0 \in [0, \infty) :$ the orbit of $x_0$ converges to 0. $\}$

(b) $\{x_0 \in [0, \infty) :$ the orbit of $x_0$ converges to 1. $\}$

(c) $\{x_0 \in [0, \infty) :$ the orbit of $x_0$ diverges to $\infty$. $\}$

4.1.3. Graphs of id and of cos are shown on $[0, \pi/2]$ in Figure 4.1.6. Does cos have a fixed point?

Let $x_0 = 0$.

(a) Draw a cobweb diagram for the iterates of $x_0$ under cos.

(b) Is the sequence of iterates monotonic? Does the sequence of iterates converge? If so, what can you say about the limit?

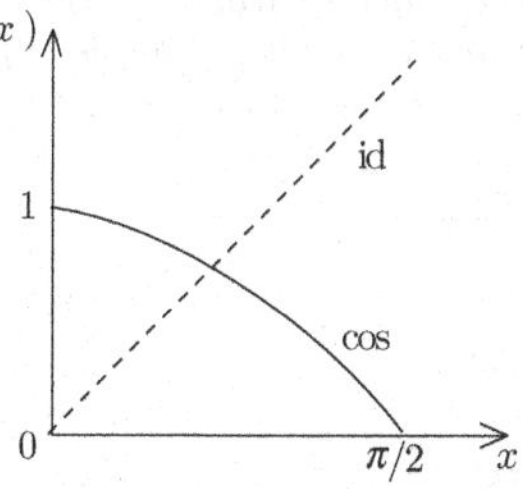


**Fig. 4.1.6** Graphs of cosine and identity mappings.

(c) Does the sequence of distances of the iterates from the limit decrease monotonically to 0?

(d) Use a pocket calculator to find the successive iterates of $x_0$ and hence find the fixed point of cos.

## 4.2 LINEAR MAPPINGS

We say that a mapping $f : \mathbb{R} \to \mathbb{R}$ is *linear* if for some $a \in \mathbb{R}$

$$f(x) = ax$$

so that $f = a\,\mathrm{id}$. The graph of $f$ is a line *through the origin* with slope $a$.

Linear mappings are the easiest types of mappings to deal with. Cobweb diagrams will be used to suggest a complete classification of their dynamics. The theory of convergence of sequences can then be used to prove what the cobweb diagrams suggest.

The dynamics of a linear mapping is determined completely by its slope $a$. The cobweb diagrams in Figures 4.2.1 show some possibilities for various values of $a$.

The origin 0 is a fixed point of the linear mapping (and if $a \neq 1$ then it is the only fixed point).

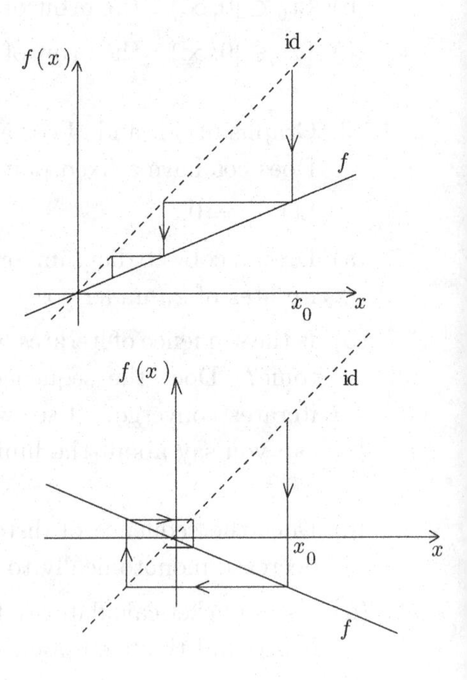


**Fig. 4.2.1(a)** $f = \frac{1}{2}\,\mathrm{id}$

The sequence of iterates converges to 0 monotonically. This is typical of the dynamics in the cases

$0 < a < 1$.

**Fig. 4.2.1(b)** $f = -\frac{1}{2}\mathrm{id}$

The sequence of iterates converges to 0 while oscillating about 0.

The distances of the iterates from 0 decrease monotonically.

This is typical of the dynamics when

$-1 < a < 0$.

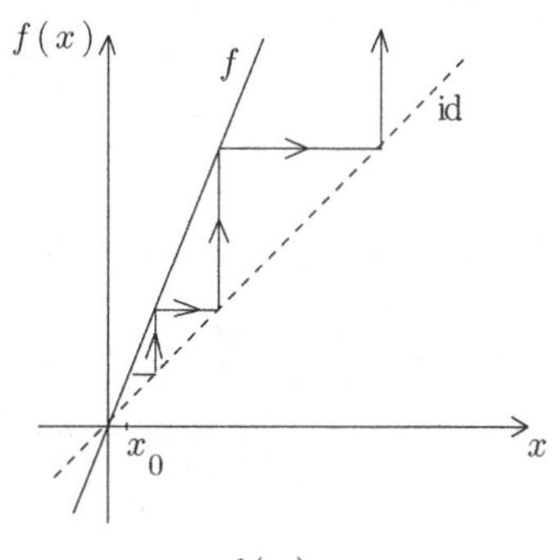


**Fig. 4.2.1(c)** $f = 2\,\mathrm{id}$
The sequence of iterates diverges to $\infty$ monotonically. This is typical of the dynamics when

$$1 < a.$$

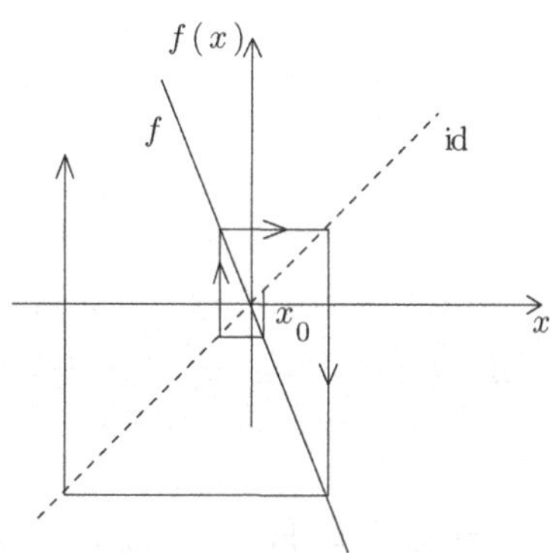


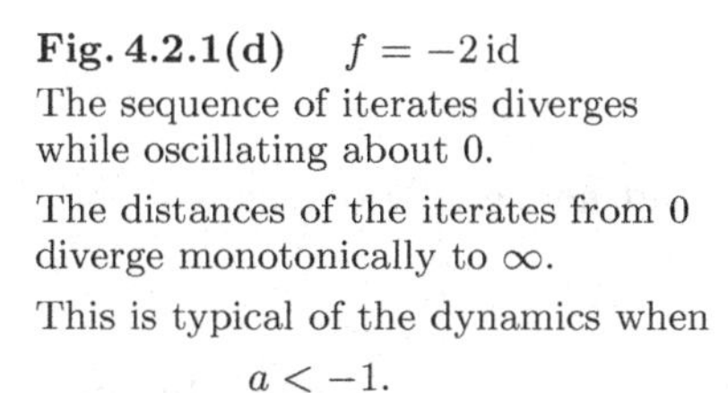
**Fig. 4.2.1(d)** $f = -2\,\mathrm{id}$
The sequence of iterates diverges while oscillating about 0.
The distances of the iterates from 0 diverge monotonically to $\infty$.
This is typical of the dynamics when

$$a < -1.$$

The Figures 4.2.1 provide an overview of the dynamics of linear mappings and suggest the following theorem:

**4.2.1 Theorem (dynamics of linear mappings)**

*Let $f : \mathbb{R} \to \mathbb{R}$ be a linear mapping with slope $a$ such that $|a| \neq 1$. For the orbit under $f$ of an element $x_0 \in \mathbb{R}$,* either

$|a| < 1$ *and the orbit converges to the fixed point* 0, or

$|a| > 1$ *and the sequence of distances of the orbit elements from* 0 *diverges to $\infty$ (unless $x_0 = 0$, when the orbit stays at the fixed point).*

To prove the theorem we use the following result from calculus.

$$\textit{the sequence}\ \ (1, a, a^2, a^3, \ldots)\ \begin{cases} \textit{converges to } 0 \textit{ if } |a| < 1 \\ \textit{but diverges to } \infty \textit{ if } 1 < a \end{cases}$$

*Proof:* Let $f$ be the linear mapping with slope $a$, so that $f(x) = ax$ and let $(x_0, x_1, x_2, \dots)$ be the orbit of $x_0$ under $f$. Hence, for all $k \in \mathbb{N}$,

$$x_{k+1} = ax_k.$$

Applying this result with $k = 0, 1, 2, \dots, n$ gives

$$\begin{aligned} x_1 &= ax_0, \\ x_2 &= ax_1 = a^2 x_0, \\ x_3 &= ax_2 = a^3 x_0, \\ &\vdots \\ x_n &= ax_{n-1} = a^n x_0. \end{aligned}$$

Here $x_0$ is a constant (independent of $n$) and so

if $|a| < 1$ then $|a|^n \to 0$, and hence $x_n \to 0$, as $n \to \infty$,

if $|a| > 1$ then $|a|^n$ diverges to $\infty$, and hence $|x_n|$ diverges to $\infty$, as $n \to \infty$ provided $x_0 \neq 0$. ■

Other details concerning the orbit $(x_0, x_1, x_2, \dots)$ follow directly from the definition

$$\begin{aligned} x_{n+1} &= f(x_n) \\ &= ax_n. \end{aligned}$$

Thus if $a \geq 0$ then the orbit is *monotonic* but if $a < 0$ then the orbit *oscillates* about the fixed point 0.

To complete our classification of the dynamics of linear mappings from $\mathbb{R}$ to $\mathbb{R}$ we note that,

*if* $a = 1$ *then every point has prime period* 1,

*if* $a = -1$ *then every point (except* 0*) has prime period-*2.

## Exercises 4.2

4.2.1. Sketch cobweb diagrams for each of the four mappings used in Figure 4.2.1 in the text, but this time suppose that the initial value for the iterates satisfies $x_0 < 0$. What happens if $x_0 = 0$?

4.2.2. Let $f : \mathbb{R} \to \mathbb{R}$ with $f(x) = -x$. Draw the cobweb diagram for the point $x_0 = 1$ and then describe the behaviour of the sequence of iterates, under $f$, of each $x \in \mathbb{R}$.

4.2.3. Let $f : \mathbb{R} \to \mathbb{R}$ with $f(x) = ax$. Prove by induction that the $n$th iterate of $x_0$ is $a^n x_0$.

4.2.4. Give an example of each of the following:

(a) A linear mapping which has only one fixed point.

(b) A linear mapping which has more than one fixed point.

(c) A linear mapping which has an orbit of prime period 2.

4.2.5. Can a linear mapping have exactly one fixed point? exactly one period-2 point?

4.2.6. Show that if $f : \mathbb{R} \to \mathbb{R}$ is linear, then $f$ has exactly one fixed point or infinitely many fixed points.

4.2.7. If a linear mapping $f$ has slope $a$, what is the slope of its second iterate $f^2$? If a linear mapping has a point with prime period 2, what can be said about $f^2$ ? about $f^3$?

4.2.8. If the cobweb diagram is a spiral, does it always go clockwise? Give reasons for your answer.

4.2.9. Let $f : \mathbb{R} \to \mathbb{R}$ be linear and let $(x_0, x_1, x_2, \ldots)$ be an orbit of $f$.

(a) Prove that $(cx_0, cx_1, cx_2, \ldots)$ is also an orbit of $f$, for each real number $c$.

(b) Prove that if $(y_0, y_1, y_0, \ldots)$ is also an orbit of $f$ then so also is $(x_0 + y_0, x_1 + y_1, x_2 + y_2, \ldots)$.

4.2.10. Prove that each linear $f : \mathbb{R} \to \mathbb{R}$ has only periodic points with prime period either 1 or 2.

[Hint. Show that if $x$ is a point with prime period $n$, then $a^n x = x$ where $a$ is the slope of the mapping. Deduce that either $x = 0$ or $a = 1$ or $-1$.]

## 4.3 AFFINE MAPPINGS

A mapping $f : \mathbb{R} \to \mathbb{R}$ is called *affine* if for some $a$ and $b$ in $\mathbb{R}$

$$f(x) = ax + b;$$

that is, $f = a\,\mathrm{id}+b$. In particular, every linear mapping is affine. An affine mapping is one which has a line (not necessarily through the origin) as its graph. The number $a$ is the *slope* of the affine mapping, which is the same as the slope of its *linear part,* $a\,\mathrm{id}$.

By elementary geometry : if $a \neq 1$, as we assume for the present, then the slope of the graph of $f$ differs from that of id.

The two graphs therefore intersect at a unique point $(p, p)$ in the plane. Hence the affine mapping $f$ has a unique fixed point $p$, as in Figure 4.3.1.

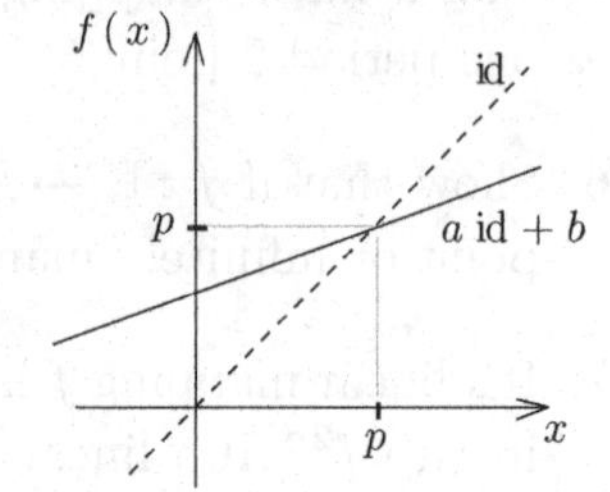


**Fig. 4.3.1** If $a \neq 1$, then $a\,\mathrm{id}+b$ has a unique fixed point $p$.

The graph of the affine mapping $a\,\mathrm{id}+b$, moreover, is obtained by translating the graph of its linear part $a\,\mathrm{id}$ through the vector $(p, p)$. Hence any cobweb diagram for the linear part gives a cobweb diagram for the affine mapping, when translated through the vector $(p, p)$, as in Figure 4.3.2.

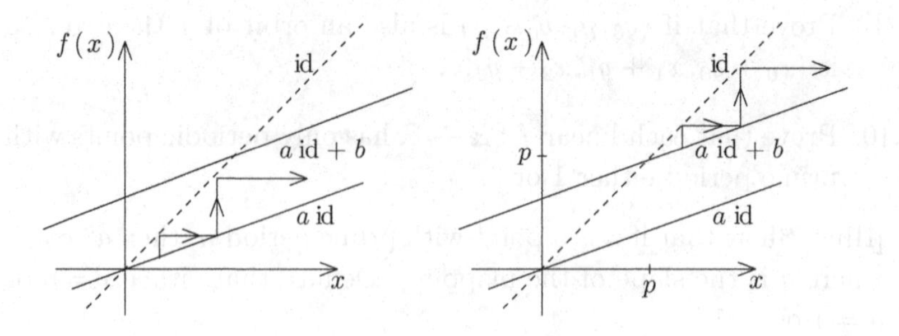


**Figure 4.3.2.** Cobweb diagram for $a\,\mathrm{id}$, translated by $(p, p)$, gives one for $a\mathrm{id}+b$.

The effect of the translation on an orbit $(x_0, x_1, x_2, \dots)$ of the linear part is to translate it along the domain axis through the distance $p$, giving the orbit $(x_0 + p,\ x_1 + p,\ x_2 + p,\ \dots)$ for the affine mapping.

Thus elementary geometry leads us to expect that the dynamics of an affine mapping will be the same as that of its linear part, but with the role of the origin now taken over by the fixed point $p$ of the affine mapping. Thus we are led to the following theorem.

### 4.3.1 Theorem (Dynamics of affine mappings)

*Let $f : \mathbb{R} \to \mathbb{R}$ be an affine mapping with slope $a$ such that $|a| \neq 1$. Let $p$ be the fixed point of the affine mapping. For the orbit under $f$ of $x_0 \in \mathbb{R}$, either*

*$|a| < 1$ and the orbit converges to the fixed point $p$, or*

*$|a| > 1$ and the sequence of distances of the orbit elements from $p$ diverges monotonically to $\infty$, provided that $x_0 \neq p$ .*

*Proof:* Since $f$ is affine, $f(x) = ax + b$ for some $b$. Now

$$\begin{aligned} x_{k+1} - p &= f(x_k) - p \\ &= f(x_k) - f(p) \\ &= ax_k + b - (ap + b) \\ &= a(x_k - p). \end{aligned}$$

Applying this with $k = 0, 1, 2, \dots$ gives

$$\begin{aligned} x_1 - p &= a(x_0 - p), \\ x_2 - p &= a(x_1 - p) = a^2(x_0 - p), \\ x_3 - p &= a(x_2 - p) = a^3(x_0 - p), \\ &\vdots \\ x_n - p &= a(x_{n-1} - p) = a^n(x_0 - p). \end{aligned}$$

Here $(x_0 - p)$ is a constant (independent of $n$); hence

if $|a| < 1$ then $a^n \to 0$, and hence $x_n \to p$ as $n \to \infty$,

if $|a| > 1$ then $|a|^n$, and hence also $|x_n - p|$, diverges monotonically to $\infty$ as $n \to \infty$. ∎

Note that the proof of Theorem 4.3.1 is the same as the proof of Theorem 4.2.1, but with $x_k$ replaced by $x_k - p$, for each $k \in \mathbb{N}$.

To complete our discussion of the dynamics of affine mappings we now consider the case in which $|a| = 1$; hence either

> $a = 1$ *and there are two cases:* $b \neq 0$ *when the affine mapping has no fixed points, and* $b = 0$ *when it has every real number as a fixed point;* or
>
> $a = -1$ *and every point, other than the fixed point* $p$*, has prime period* 2.

## Asymptotic behaviour

In the examples of orbits of linear mappings shown in the Figures 4.2.1, some orbits converge to a fixed point, while others move away from it. In the former case we say that the orbit is *asymptotic* to the fixed point, while in the latter case we say that the orbit is *backwards asymptotic* to the fixed point.

Note that in Figures 4.2.1(c) and (d), the orbit of $x_0$ can be continued backwards by simply using the cobweb construction in reverse (that is, leave the graph of id horizontally and the graph of $f$ vertically.) In this way we get a 'backwards orbit' of $x_0$ which converges to the fixed point. This is the idea behind the phrase 'backwards asymptotic'.

## Exercises 4.3

4.3.1. Are all linear mappings affine? Give an example of an affine mapping which is not linear.

4.3.2. What are the possible numbers of fixed points for (i) a linear mapping from $\mathbb{R}$ to $\mathbb{R}$? (ii) an affine mapping from $\mathbb{R}$ to $\mathbb{R}$?

4.3.3. Verify algebraically that if $a \neq 1$ then the affine mapping $a\,\mathrm{id}+b$ has just one fixed point $p$, given by the formula $p = \frac{b}{1-a}$.

4.3.4. Use the cobweb diagram for the linear mapping $2\,\mathrm{id}$ shown in Figure 4.2.1(c) to obtain a cobweb diagram for the affine mapping $2\,\mathrm{id} - 1$.

4.3.5. Let $(x_0, x_1, x_2, \dots)$ be an orbit of the affine mapping $a\,\mathrm{id} + b$ with $a \neq 1$ and with $p$ as fixed point. Use Mathematical Induction to prove carefully that $x_n - p = a^n(x_0 - p)$, for all $n \in \mathbb{N}$.

---

## Additional reading for Chapter 4

The use of cobweb diagrams is called 'graphical analysis' in Section 3.2 of [De2], Section 1.3 of [De1], Chapter 1 of [Gul] and in Section 4.1 of [Ho]. It is called 'graphical iteration' in Section 4.1 of [PMJPSY] and in Section 1.5 of [PJS].

Most authors do not treat the dynamics of linear and affine maps on $\mathbb{R}$ separately, preferring to consider differentiable maps in general. We study differentiable maps in the next chapter.

---

## References

[De1] Robert L. Devaney, *An Introduction to Chaotic Dynamical Systems: Second Edition,* Addison-Wesley, Menlo Park California, 1989.

[De2] Robert L. Devaney, *Chaos, Fractals and Dynamics, Computer Experiments in Mathematics,* Addison-Wesley, New York, 1990.

[Gul] Denny Gulick, *Encounters with Chaos,* McGraw-Hill, Inc., 1992.

[Ho] Richard Holmgren, *A First Course in Discrete Dynamical Systems* Springer-Verlag, 1994.

[PMJPSY] Hans-Otto Peitgen, Evan Maletsky, Hartmut Jürgens, Terry Perciante, Dietmar Saupe, and Lee Yunker, *Fractals for the Classroom: Strategic Activities Volume Two,* Springer-Verlag, 1992. (two volumes)

[PJS] Hans-Otto Peitgen, Hartmut Jürgens and Dietmar Saupe, *Chaos and Fractals, New Frontiers of Science*, Springer-Verlag, New York, 1992.

# 5

# ASYMPTOTIC ORBITS II
## Differentiable mappings

*The main idea in this chapter is very simple. For a mapping which is differentiable, the graph has a tangent at each point. Near the point of tangency the graph stays very close to the tangent. But the tangent is the graph of an affine mapping and so the dynamics of the differentiable map should be close to that of the affine mapping. We use this to predict the dynamics of a differentiable mapping near a fixed point.*

*We begin by using the idea of zooming to help us express the ideas of tangency and differentiability in terms of modern computer graphics. Graphs are then used to motivate the main theorem on the dynamics of a differentiable mapping near a fixed point. This leads us to the ideas of attracting, repelling and indifferent fixed points.*

*Finally, we use the results for dynamics of mappings near fixed points to study their dynamics near periodic points and orbits.*

## 5.1 DIFFERENTIABLE MAPPINGS

In this section we show how tangents to curves can be obtained by zooming. We then show the relevance of this to the dynamics of differentiable mappings near their fixed points.

### Tangents

The problem of finding a tangent at some point on a curve is the geometric motivation for the study of differential calculus. Beginners in the subject are assumed to have an intuitive understanding of the idea of a tangent. Derivatives are introduced as a means for calculating the slope of the tangent. We shall go along with this approach, while giving extra emphasis to the importance of the tangents themselves (as distinct from their slopes).

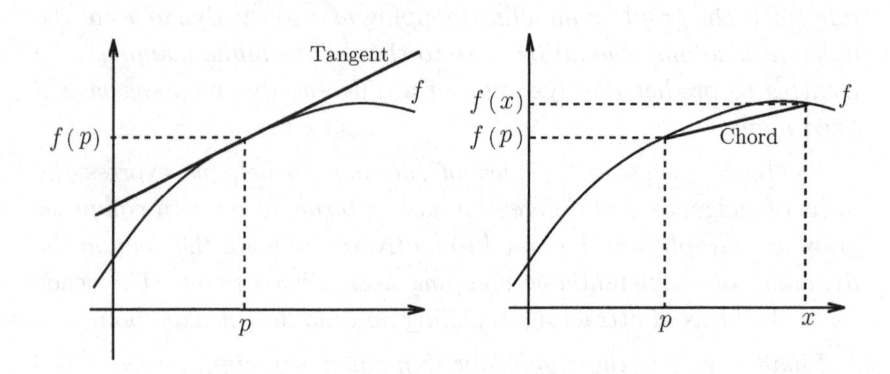


A tangent to graph of $f$ A typical chord on the graph.

**Fig. 5.1.1**

Recall that the ratio

$$\frac{f(x) - f(p)}{x - p}$$

is the slope of the line segment joining the points $(x, f(x))$ and $(p, f(p))$ on the graph of $f$. We regard $p$ as fixed and then call this ratio the *Newton quotient* of $f$ at $x$.

Figure 5.1.1 suggests that as $x$ approaches $p$ the Newton quotient will approach the slope of the tangent to the graph of $f$ at the point $(p, f(p))$. This is the idea behind the following definition:

> Let $f : S \to S$ where $S$ is a subinterval of $\mathbb{R}$. The mapping $f$ is *differentiable* at $p$ if and only if the following limit exists
>
> $$\lim_{x \to p} \frac{f(x) - f(p)}{x - p}.$$
>
> This limit is denoted by $f'(p)$ and called the *derivative* of $f$ at $p$

Given this definition, it is possible to make the idea of a tangent more precise: we say that the graph of a function $f$ has a (non-vertical) tangent at the point $(p, f(p))$ if and only if $f$ is differentiable at $p$. The tangent is then the line through the point $(p, f(p))$ with slope $f'(p)$.

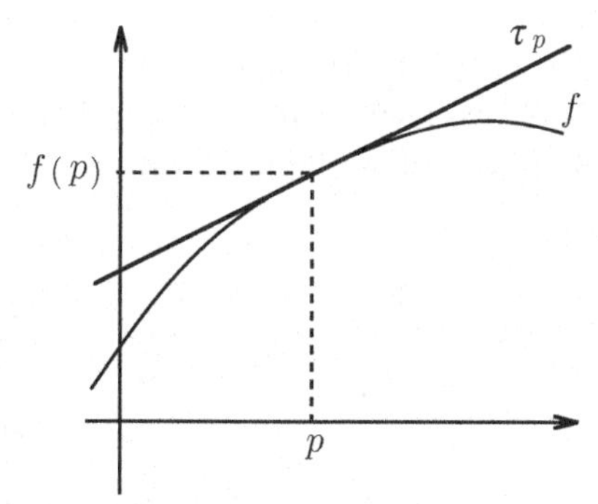


**Fig. 5.1.2** Graph of $\tau_p$ is the tangent to the graph of $f$ at $(p, f(p))$.

Since we want to emphasize the tangent itself rather than just its slope, we define the *tangent mapping* at $p$, $\tau_p : \mathbb{R} \to \mathbb{R}$, by putting

$$\tau_p(x) = f(p) + f'(p)(x - p). \tag{1}$$

The mapping $\tau_p$ is affine, and also $\tau_p(p) = f(p)$ and $\tau_p'(p) = f'(p)$. Hence the graph of $\tau_p$ is the straight line passing through the point $(p, f(p))$ with slope $f'(p)$. The graph of $\tau_p$ is thus the tangent to the graph of $f$ at the point $(p, f(p))$, as in Figure 5.1.2.

## Tangents via zooming

We can now relate the idea of tangency to that of zooming in on a point on the graph. We shall need the following lemma.

**5.1.1 Lemma** *Let $f : S \to S$ be differentiable at $p \in S$. If $\tau_p$ is the tangent mapping for $f$ at $p$, then*

$$\lim_{x \to p} \frac{1}{|x-p|} \, |f(x) - \tau_p(x)| = 0. \tag{2}$$

*Proof:* This is set as Exercise 5.1.5 (and is essentially just a rewrite of the definition of the derivative and of the tangent mapping $\tau_p$). ∎

Now (2) says, intuitively, that the product

$$\frac{1}{|x-p|} \, |f(x) - \tau_p(x)|$$

gets very small when $x$ gets very close to $p$. Hence the discrepancy $|f(x) - \tau_p(x)|$ between the mappings $f$ and $\tau_p$ not only gets small, but it still gets small even when it is multiplied by the large factor $1/|x-p|$. The geometrical interpretation of this idea is shown in Figure 5.1.3.

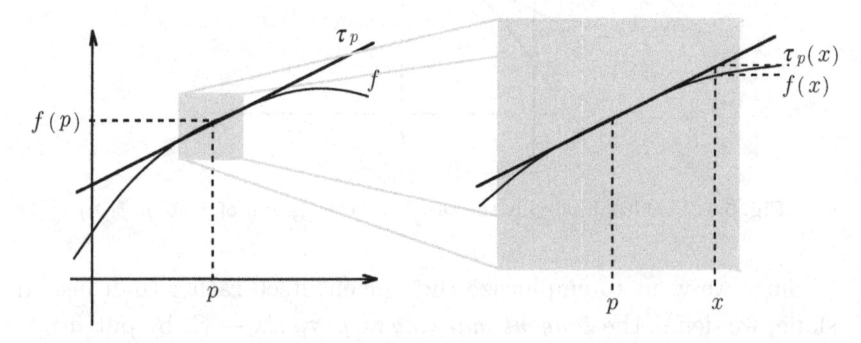


**Fig. 5.1.3.** The small square on the left containing the point $p$ is magnified by the factor $1/|x-p|$ to have sides of unit length. The magnified square is shown on the right. Even after magnification, the graphs of $f$ and $\tau_p$ stay close.

This can all be expressed very simply if we refer to the zoom facility on a computer, which allows us to select a small portion of the display on the screen and then magnify it so as to fill the entire screen:

*If we repeatedly zoom in on a point of the graph of a differentiable function, then the magnified portion of the graph approaches (a segment of) the tangent to the graph at the point.*

## Zooming near a fixed point

Figure 5.1.4 illustrates this. It shows the graph of a differentiable mapping $f$ which has a fixed point 0. It also shows the graph of an orbit of $f$. This orbit is the one with initial condition $x_0$ where $x_0 \approx \frac{1}{2}$.

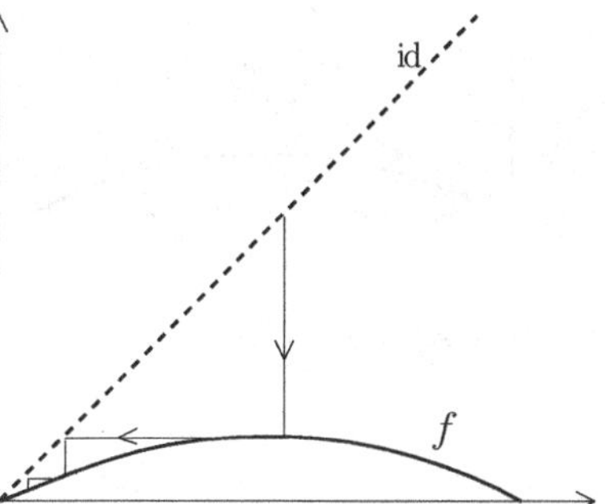


**Figure 5.1.4** $f(x) = 0.5x(1 - x)$.

This mapping has a fixed point at 0 and the orbit of $x_0$ converges to the fixed point monotonically.

**5.1.2 Example** *Let $f$ be the differentiable mapping shown in Figure 5.1.4.*

*By zooming in on this fixed point, illustrate the idea that the dynamics of the mapping is like that of its tangent mapping (while the orbit stays close to the fixed point).*

*Solution:*

On the left of Figure 5.1.5 smaller and smaller squares are shown, each having a corner at the origin. On the right we magnify each of these squares by factors which bring them to a common size. As the magnification increases, two things happen:

(a) the *magnified graphs* approach (a segment of) the tangent to the graph of $f$ at the origin;

(b) the *magnified cobweb diagram* seems to be approaching the sort of cobweb diagram we would expect for the tangent mapping $\frac{1}{2}\text{id}$.

Thus Figure 5.1.5 suggests that near the fixed point 0, the affine mapping $\frac{1}{2}\text{id}$ is a good approximation to the differentiable mapping $f$, and the dynamics of the two mappings are very similar.

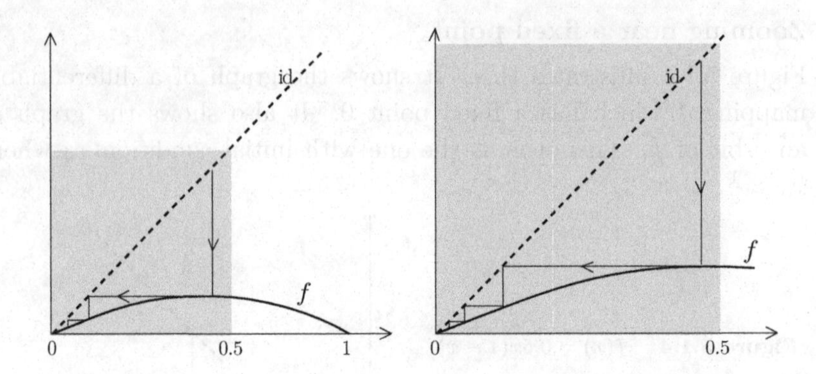


**(a)** The square on the left magnified by 2 gives the square on the right.

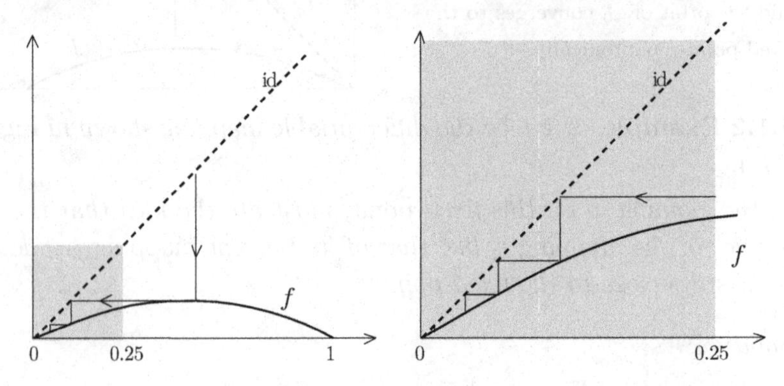


**(b)** The square on the left magnified by 4 gives the square on the right.

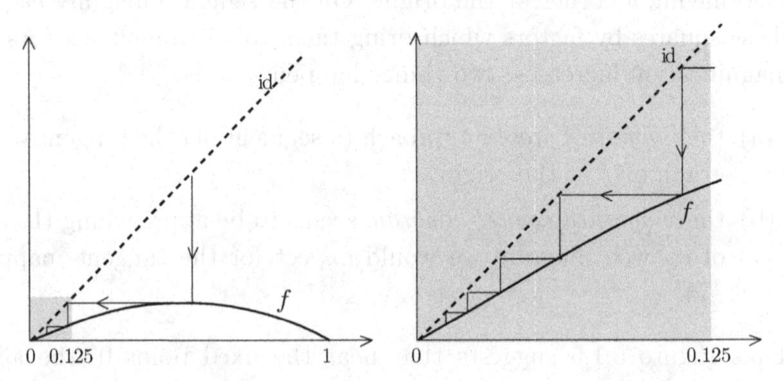


**(c)** The square on the left magnified by 8 gives the square on the right.

**Figure 5.1.5.** As the magnification is increased, the quadratic mapping near 0 looks more and more like an affine mapping. ▪

## Attractors and repellors

The above discussion suggests that *near a fixed point* $p$, the dynamics of a differentiable mapping $f$ should be like those of the tangent mapping at $p$, which is *affine* with slope $f'(p)$.

Hence by Theorem 4.3.1, the possibilities for the cobweb diagrams for $f$ near the fixed point $p$, should be as in Figure 5.1.6:

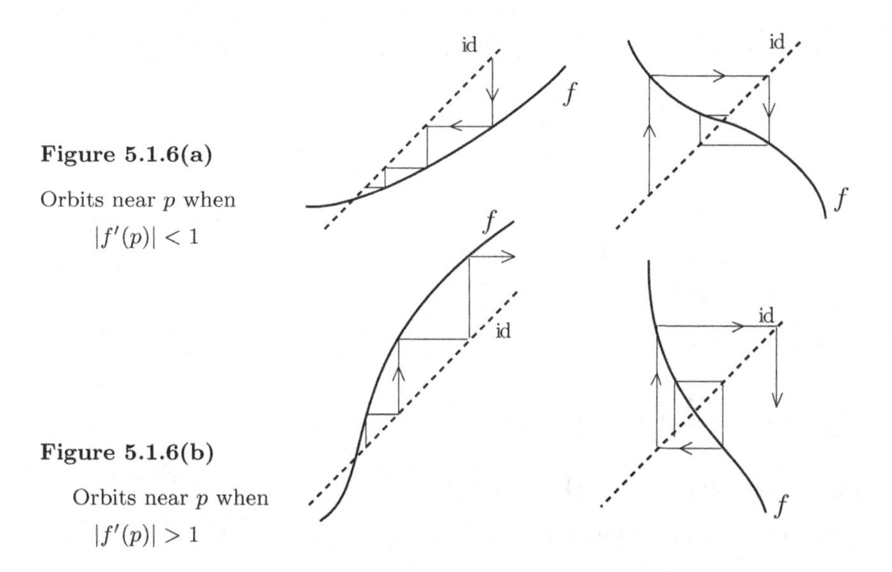


**Figure 5.1.6(a)**

Orbits near $p$ when $|f'(p)| < 1$

**Figure 5.1.6(b)**

Orbits near $p$ when $|f'(p)| > 1$

Thus the behaviour of the orbits of a mapping $f$ near a fixed point $p$ should be determined by the magnitude of the derivative $|f'(p)|$. Hence we make the following definition.

**5.1.3 Definition** Let $f : I \to \mathbb{R}$, where $I$ is an interval. A fixed point $p$ of a mapping is called

an *attractor* (or an *attracting fixed point*) if $|f'(p)| < 1$,

a *repellor* (or a *repelling fixed point*) if $|f'(p)| > 1$, and

an *indifferent* fixed point if $|f'(p)| = 1$.

The number $f'(p)$ is called the *multiplier* of the fixed point $p$. ∎

Attractors should attract nearby orbits and repellors should repel them. That this actually happens will be proved in the next section.

Indifferent fixed points can show attracting or repelling behaviour, or a combination of both types. The real significance of indifferent fixed points, however, will not be revealed until later, in the study of bifurcation of fixed points.

A fixed point which is an attractor is said to be *stable*, and one which is a repellor is said to be *unstable*.

## Exercises 5.1

5.1.1. Let $f : \mathbb{R} \to \mathbb{R}$ with $f(x) = x^2$ and let $p \in \mathbb{R}$. Find the tangent mapping $\tau_p$ to $f$ at $p$.

5.1.2. For each of the following mappings $f : [0,1] \to [0,1]$ find all the fixed points and state whether they are attracting, repelling or indifferent.

(a) $f(x) = 2x(1-x)$,

(b) $f(x) = \frac{1}{2}x(1-x)$,

(c) $f(x) = 2x^2(1-x)$.

5.1.3. Let $f : \mathbb{R} \to \mathbb{R}$ with $f(x) = \frac{1}{2}(x^2 + 1)$.

(a) Prove that $f$ has only one fixed point and that it is an indifferent fixed point.

(b) Show that there are points arbitrarily close to the fixed point whose orbits converge to the fixed point

(c) Show that there are points arbitrarily close to the fixed point whose orbits diverge away from the fixed point.

5.1.4. Let $f : \mathbb{R} \to \mathbb{R}$ be an affine mapping. Show that the tangent mapping for $f$ at each $p \in \mathbb{R}$ is equal to $f$.

5.1.5. Prove Lemma 5.1.1.

## 5.2 MAIN THEOREM AND PROOF

The following theorem is our main result about the behaviour of a mapping near a fixed point. It is the culmination of the previous chapter and the current chapter.

A new feature is the use of an interval $I$, whose purpose is to measure closeness to $p$.

By saying that the interval $I$ is *open-in-$S$* we mean that it is the intersection of $S$ with an open interval.

### 5.2.1 Theorem (Dynamics near a fixed point)

*Let$f : S \to S$, where $S$ is an interval. Suppose that $p \in S$ is a fixed point of $f$ and that $f$ is differentiable at $p$ with $|f'(p)| \neq 1$.*

*Case (i): $|f'(p)| < 1$. There is an open-in-$S$ interval $I$ containing $p$ such that every orbit $(x_0, x_1, x_2, \dots)$ with initial point $x_0 \in I$ converges to the fixed point $p$. If $x_n \neq p$, then the next point $x_{n+1}$ of the orbit is closer to $p$ than $x_n$ is.*

*Case (ii): $|f'(p)| > 1$. There is an open-in-$S$ interval $I$ containing $p$ such that no orbit with initial point $x_0 \in I \setminus \{p\}$ remains in $I$. If $x_n \in I \setminus \{p\}$, then the next point $x_{n+1}$ of the orbit is further from $p$ than $x_n$ is.*

In this section, we prove Theorem 5.2.1 after introducing a preliminary result on differentiable mappings (Lemma 5.2.2).

The proof of Theorem 5.2.1 for differentiable mappings is similar to that of Theorem 4.3.1 for affine mappings. Now, however, the mapping $f$ is only *approximately* affine near the fixed point $p$ and hence the key equalities in the earlier proofs are replaced here by inequalities.

**5.2.2 Lemma** *Suppose $f : S \to S$ is differentiable at $p \in S$ and $|f'(p)| < 1$. Then there is a positive number $a < 1$ and an open-in-$S$ interval $I$ such that, for all $x \in I$,*

$$|f(x) - f(p)| \leq a|x - p|.$$

*Proof:* Let $f$ be differentiable at $p$. The definition of $f'(p)$ as the limit of a Newton quotient gives

$$\lim_{x \to p} \frac{f(x) - f(p)}{x - p} = f'(p). \tag{1}$$

Suppose, furthermore, that $|f'(p)| < 1$. Choose a number $a$ such that $|f'(p)| < a < 1$ and hence $-a < f'(p) < a$ and hence

$$f'(p) \text{ is in the interval } (-a, a). \tag{2}$$

By the definition of limit, *there is therefore an interval $I$ containing $p$ such that for all $x \in I$ with $x \neq p$, the value of Newton quotient at $x$ is in the interval* $(-a, a)$. (see Figure 5.2.1)

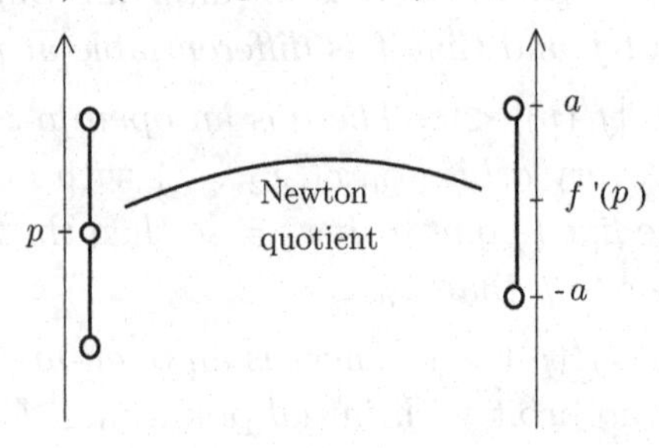


**Figure 5.2.1.** How the Newton quotient maps $I \setminus \{p\}$.

The italicized statement implies that, for all $x \in I$ with $x \neq p$,

$$-a < \frac{f(x) - f(p)}{x - p} < a$$

and hence that

$$\left| \frac{f(x) - f(p)}{x - p} \right| < a.$$

Thus, for all $x \in I$,

$$|f(x) - f(p)| \leq a|x - p|. \qquad \blacksquare$$

**Remark.** In cases where $p$ is not an endpoint of the interval $S$, we may suppose, by taking a smaller interval if necessary, that the interval $I$ occurring in Lemma 5.2.2 has its centre at $p$. Hence, *if a point is in $I$, so is any point at a smaller distance from $p$.*

*Proof of Theorem 5.2.1:* Suppose first that $|f'(p)| < 1$.

Now since $p$ is a fixed point of $f$ we deduce from Lemma 5.2.2 that for each $x \in I$,

$$\big|f(x) - p\big| \leq a\big|x - p\big|.$$

If $x_k \in I$ we may substitute $x_k$ in place of $x$ and $x_{k+1}$ in place of $f(x)$ to get

$$|x_{k+1} - p| \leq a\,|(x_k - p)|. \tag{3}$$

Hence also $x_{k+1} \in I$ since $a < 1$.

By hypothesis $x_0 \in I$. Hence we may apply (1) with $k = 0, 1, 2, \ldots$ to get

$$\begin{aligned}
|x_1 - p| &\leq a\,|x_0 - p|, \\
|x_2 - p| &\leq a\,|x_1 - p| \quad \leq a^2|x_0 - p|, \\
|x_3 - p| &\leq a\,|x_2 - p| \quad \leq a^3|x_0 - p|, \\
&\vdots \\
|x_n - p| &\leq a\,|x_{n-1} - p| \leq a^n|x_0 - p|.
\end{aligned}$$

But $0 \leq a < 1$ so that $a^n \to 0$ as $n \to \infty$.

Since $|x_0 - p|$ is a constant (independent of $n$), it follows that

$$\lim_{n\to\infty} |x_n - p| = 0$$

and so

$$\lim_{n\to\infty} x_n = p \tag{4}$$

Finally, the proof when $|f'(p)| > 1$ is left as Exercise 5.2.2. ∎

## Exercises 5.2

5.2.1. Modify Lemma 5.2.2 when $|f'(p)| > 1$ and then prove it.

5.2.2. Prove Theorem 5.2.1 when $|f'(p)| > 1$.

## 5.3 PERIODIC POINTS

The theory in the previous sections has been concerned with the orbits which are asymptotic to a *fixed point*. We now show how it can be used to study orbits which are asymptotic to an orbit of period $n$.

To illustrate the problem, we show in Figures 5.3.1 a period-2 orbit of a mapping $f$ and then some orbits which are asymptotic to it.

**Figure 5.3.1(a)** $f(x) = 3.2x(1 - x)$

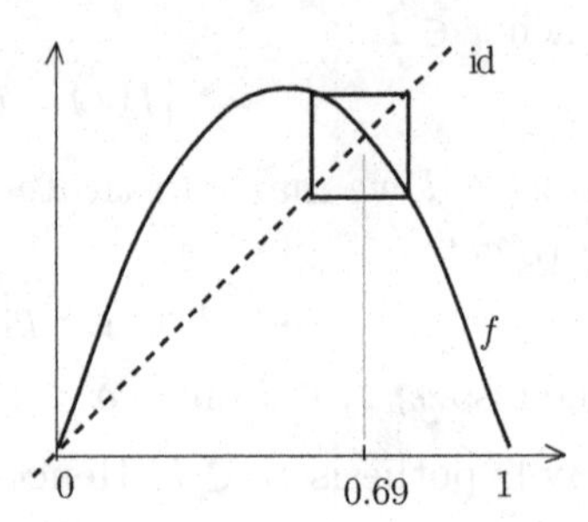


The cobweb diagram shows the square shape which is characteristic of period-2 orbits.

The mapping $f$ also has two fixed points. The first is at 0 and the second is approximately 0.69.

**Figure 5.3.1(b)**

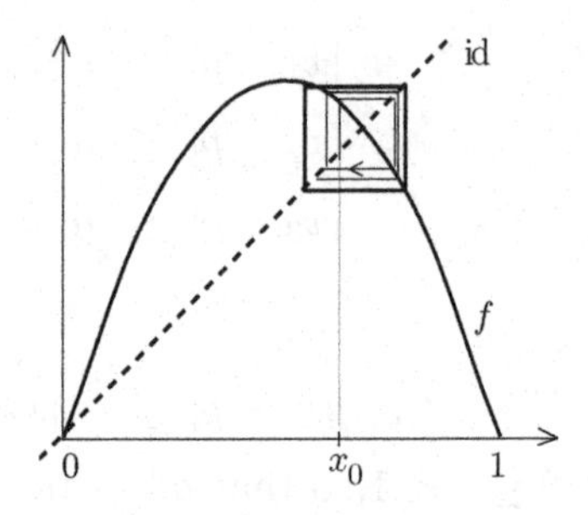


The cobweb which starts inside the period-2 square spirals out towards it.

Thus the orbit of $x_0$ is backwards asymptotic to the second fixed point and is asymptotic to the period-2 orbit.

**Figure 5.3.1(c)**

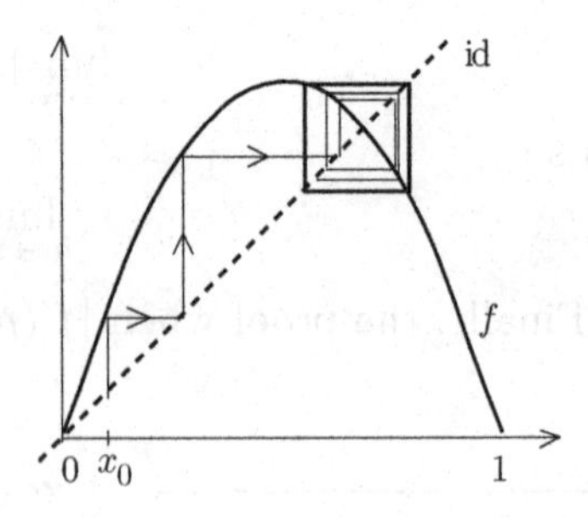


The cobweb moves away from the origin monotonically. Eventually it gets inside the period-2 square and thereafter it spirals out towards the square.

Thus the orbit of $x_0$ is backwards asymptotic to the first fixed point and is asymptotic to the period-2 orbit.

To obtain a more detailed description of the spiralling in Figures 5.3.1, we magnify the period-2 square and the spiral which it contains. We then relate the spiralling to what is happening on the graph of $f^2$ where the orbits are asymptotic to *fixed points* of $f^2$ instead of *prime period-2 points* of $f$.

The two prime period-2 points of $f$ are denoted by $a_0$ and $a_1$. The orbit of $x_0$, which is asymptotic to the prime period-2 orbit of $f$, is denoted by

$$(x_0, x_1, x_2, \ldots).$$

**Figure 5.3.2** $f(x) = 3.2x(1-x)$

Zooming in on the period-2 square from Figure 5.3.1 gives the enlarged cobweb for the asymptotic orbit.

The orbit of $x_0$ can be split into two sequences:

$$(x_0, x_2, x_4, \ldots)$$

converging to the period-2 point $a_0$

and

$$(x_1, x_3, x_5, \ldots)$$

converging to the period-2 point $a_1$.

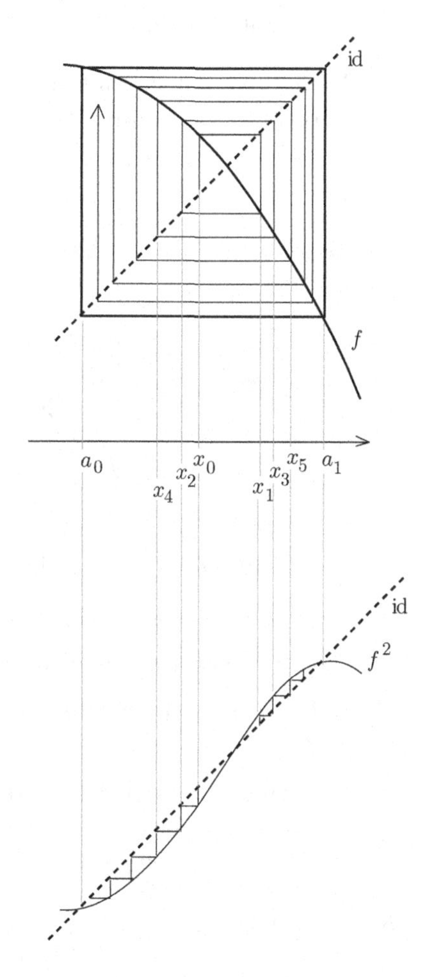


Part of the common domain for each of the mappings $f$ and $f^2$, containing the interval $[a_0, a_1]$, is shown.

The part of the graph of $f^2$ which lies above the interval $[a_0, a_1]$ is shown.

The sequence

$$(x_0, x_2, x_4, \ldots)$$

converges monotonically to the fixed point $a_0$ of $f^2$ while the sequence

$$(x_1, x_3, x_5, \ldots)$$

converges monotonically to the fixed point $a_1$.

## Exercises 5.3

5.3.1. For the mapping $f : [0,1] \to [0,1]$ with $f(x) = 3.2x(1-x)$ (from Figure 5.3.1) sketch the cobweb diagram for an orbit which is
(i) backwards asymptotic to the fixed point 0, and
(ii) eventually equal to the other period-1 orbit.

5.3.2. Repeat the previous exercise, but with (ii) replaced by the condition that the orbit is eventually equal to the prime period-2 orbit (shown in Figure 5.3.1).

5.3.3. Figure 5.3.3 shows the cobweb diagram for an orbit which appears to be asymptotic to a prime period-2 orbit.

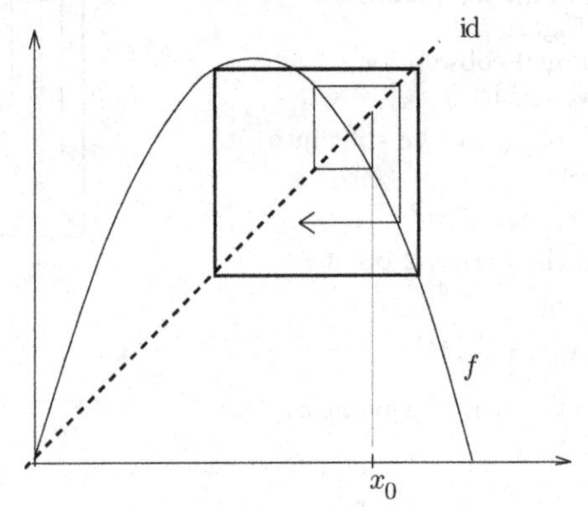


**Figure 5.3.3.** $f : [0,1] \to [0,1]$ with $f(x) = 3.45x(1-x)$

(a) Does the cobweb ever leave the box?

(b) If it leaves, does it ever get back inside the box?

(c) How can you determine the asymptotic behaviour of the orbit ?

(d) Sketch a cobweb for an orbit which is backwards asymptotic to the nonzero fixed point of $f$ and which is eventually equal to the period-2 orbit.

5.3.4. Plot the graph of $f : [-1,1] \to [-1,1]$ with $f(x) = \sqrt{1-x^2}$ (the upper unit semicircle). Which point is the fixed point of $f$ ? Draw cobweb diagrams for different starting values in the interval $[-1,1]$ and discuss the behaviour of the orbits.

5.3.5. Let $f : S \to S$ where $S$ is a set and let $x_0 \in S$. Show that
if $(x_0, x_1, x_2, \dots)$ is the orbit of $x_0$ under $f$,
then $(x_0, x_2, x_4, \dots)$ is the orbit of $x_0$ under $f^2$ and
$(x_1, x_3, x_5, \dots)$ is the orbit of $x_1$ under $f^2$.

5.3.6. State the converse of the result in the previous exercise. Show that the statement you have written down is true if $x_1 = f(x_0)$.

5.3.7. Show that
if $(x_0, x_1, x_2, x_3, x_4, x_5, x_6, x_7, x_8, \dots)$ is the orbit of $x_0$ under $f$,
then $(x_0, x_3, x_6, \dots)$ is the orbit of $x_0$ under $f^3$,
$(x_1, x_4, x_7, \dots)$ is the orbit of $x_1$ under $f^3$ and
$(x_2, x_5, x_8, \dots)$ is the orbit of $x_2$ under $f^3$.

5.3.8. State the converse of the result of the previous exercise. Give a sufficient condition for the statement you have written down to be true.

---

## Additional reading for Chapter 5

Section 1.2 of [Gul] takes a different approach to defining attracting and repelling fixed points. Instead of using derivatives, Gulick considers directly whether nearby points move toward or away from the fixed point. We call this a *topological* approach. He gives a proof of our main theorem in this context. He takes a similar approach to attracting and repelling periodic orbits in Section 1.3.

Devaney gives a topological definition of attracting and repelling periodic orbits in Section 3.3 of [De2], but section 1.4 of [De1], gives the definitions using derivatives that we have seen in this chapter.

There is also an intuitive discussion of attracting periodic orbits without any definitions in Section 4.3 of [PMJPSY].

## References

[De1] Robert L. Devaney, *An Introduction to Chaotic Dynamical Systems: Second Edition,* Addison-Wesley, Menlo Park California, 1989.

[De2] Robert L. Devaney, *Chaos, Fractals and Dynamics, Computer Experiments in Mathematics,* Addison-Wesley, New York, 1990.

[Gul] Denny Gulick, *Encounters with Chaos,* McGraw-Hill, Inc., 1992.

[PMJPSY] Hans-Otto Peitgen, Evan Maletsky, Hartmut Jürgens, Terry Perciante, Dietmar Saupe, and Lee Yunker, *Fractals for the Classroom: Strategic Activities Volume Two,* Springer-Verlag, 1992. (two volumes)

# 6

# FAMILIES OF MAPPINGS AND BIFURCATION

*The previous chapters are mainly about the dynamics of individual mappings. There are important questions in dynamical systems, however, which are expressed in terms of a family of mappings, rather than an individual mapping. The need to consider families of mappings comes, in part, from applications — such as those introduced in Chapter 1 — where the right-hand side mappings of the difference equations depended on a parameter.*

*In the context of families of mappings, we introduce a new type of diagram which shows how the fixed points change with a parameter on which the mapping depends. This leads to the concept of bifurcation of families of fixed points. The start of the famous period-doubling cascade for the logistic family of mappings is used as an illustration.*

## 6.1 THREE FAMILIES OF MAPPINGS

Let $f_\mu$ denote a mapping which depends on a parameter $\mu$. As $\mu$ varies so does the mapping $f_\mu$. This gives rise to a function $\mu \mapsto f_\mu$, which assigns a mapping $f_\mu$ to each $\mu$ in some set. We call the function $\mu \mapsto f_\mu$ a *family of mappings.*

This section provides notation and terminology for three specific families of mappings which will be used later to illustrate chaotic dynamics.

**6.1.1 Definition** The *logistic family* of mappings $\mu \mapsto Q_\mu$ is obtained by putting

$$Q_\mu(x) = \mu x(1 - x), \tag{1}$$

where $x$ and $\mu$ are to be restricted to suitable sets. ∎

Four members of the logistic family are shown in Figure 6.1.1, where the domain of each mapping is chosen as the interval $[0, 1]$.

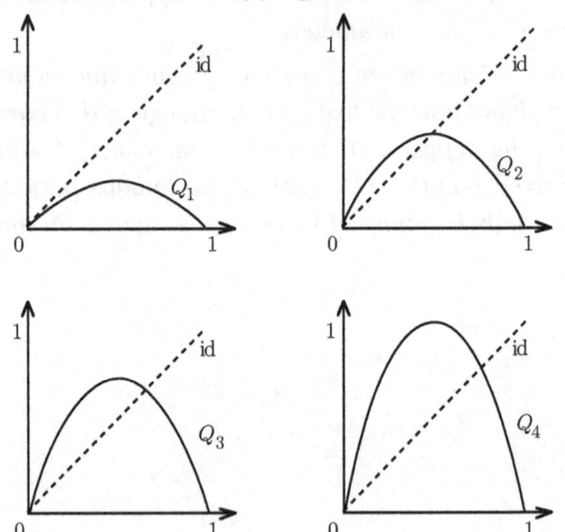


**Figure 6.1.1.** Graphs of four logistic mappings $Q_1$, $Q_2$, $Q_3$, and $Q_4$. As $\mu$ increases from 0 to 4, the maximum value of $Q_\mu$ increases from 0 to 1.

The choice of the interval $[0, 1]$ for the domain of $Q_\mu$ is suggested by the discrete logistic model of population growth in Chapter 1.

The size of the population cannot be negative, either now or after the next breeding season. Hence we require both $x \geq 0$ and $Q_\mu(x) \geq 0$. By (1) the latter condition (when $x \geq 0$ and $\mu > 0$) is just $x \leq 1$.

For $0 \leq \mu \leq 4$ it is easy to verify that $Q_\mu$ maps the interval $[0, 1]$ into itself. As $\mu$ increases, so does the complexity of the dynamics of mapping $Q_\mu$. Later we use the case $\mu = 4$ to introduce the definition of chaos. For $\mu = 2$ and for $\mu = 4$, there are closed formulae for the orbits of $Q_\mu$: Exercises 6.1.8 and 6.1.9 show how to get them.

**6.1.2 Definition** The family of *tent mappings* $\mu \mapsto T_\mu$ is obtained by putting

$$T_\mu(x) = \frac{\mu}{4}(1 - |2x - 1|).$$

with $x$ and $\mu$ restricted to suitable sets. ■

Four members of the tent family are shown in Figure 6.1.2, where once again we have chosen the domain to be the interval $[0, 1]$.

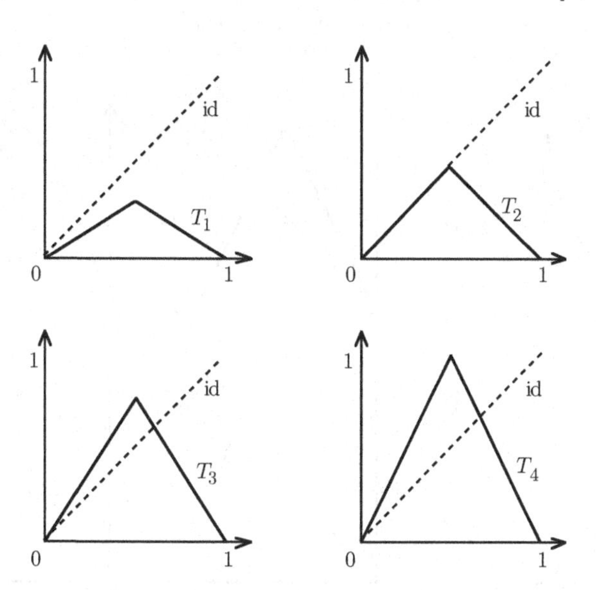


**Figure 6.1.2.** Graphs of four tent mappings $T_1$, $T_2$, $T_3$ and $T_4$. As $\mu$ increases from 0 to 4, the maximum value of $T_\mu$ increases from 0 to 1.

For $0 \leq \mu \leq 4$, it is easy to verify that $T_\mu$ maps the interval $[0, 1]$ into itself. As $\mu$ increases, so does the complexity of the dynamics of the mapping $T_\mu$. Later we use the case $\mu = 4$ to illustrate the definition of chaos.

The tent family shares many properties of the logistic family (in spite of the fact that $T_\mu$ is not differentiable when $\mu > 0$).

**6.1.3 Definition** The connecting family of mappings $\mu \mapsto C_\mu$ is obtained by putting

$$C_\mu(x) = 1 - |2x - 1|^\mu$$

where $x \in [0, 1]$ and $\mu > 0$. ∎

Typical members of the connecting family are shown in Figure 6.1.3. This family has some curious properties. It connects the logistic mapping $Q_4$ to the tent mapping $T_4$ and all its members with $\mu > \frac{1}{2}$ have the same dynamics, as will be shown later.

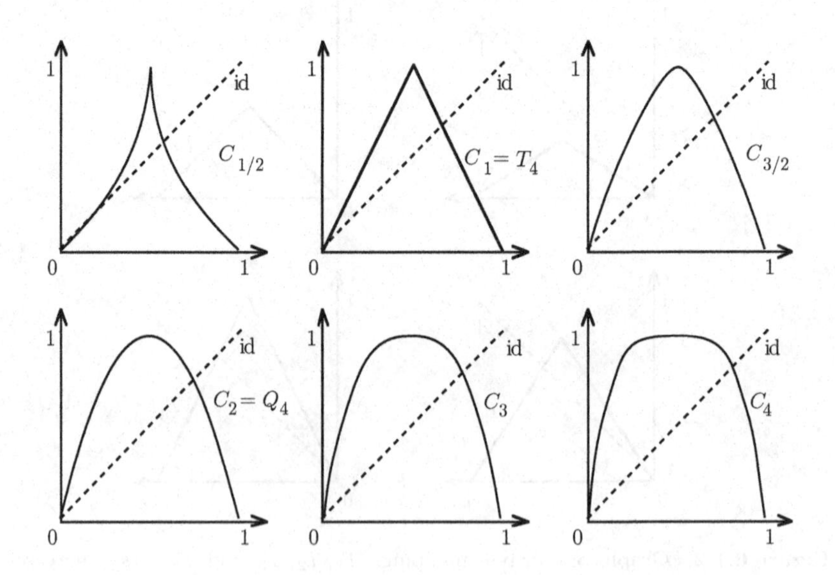


**Figure 6.1.3.** Graphs of six mappings $C_{1/2}$, $C_1$, $C_{3/2}$, $C_2$, $C_3$, $C_4$ from the connecting family. The maximum value of $C_\mu$ is 1, for each $\mu > 0$.

The three families of mappings defined above share a number of common features. To describe a couple of these features let $f$ denote a mapping in any one of these families.

First, the mapping $f$ satisfies $f(\frac{1}{2}-z) = f(\frac{1}{2}+z)$ for all $z \in [0, \frac{1}{2}]$, and so $f$ is *symmetric* about $x = \frac{1}{2}$.

Second, the values of $f$ rise steadily from 0 at the left endpoint of the domain to a maximum value and then drop back steadily to 0.

**6.1.4 Definition** A mapping $f : [0, 1] \to [0, 1]$ is called *unimodal* if it assumes a maximum value at some point $c$ where $0 < c < 1$, and is strictly monotonic on the intervals $[0, c]$ and $[c, 1]$. ■

Thus every logistic mapping, tent mapping and connecting mapping is unimodal.

## Exercises 6.1

6.1.1. Show that the mappings $Q_4$ and $C_2$ are equal, where
$$Q_4(x) = 4x(1-x) \qquad \text{and} \qquad C_2(x) = 1 - |2x-1|^2.$$

6.1.2. Show that $Q_\mu(x) = \frac{\mu}{4}\left(1 - (2x-1)^2\right)$.

6.1.3.

(a) Show that $Q_\mu(x)$ has its maximum value of $\mu/4$ at $x = \frac{1}{2}$.

(b) Does $Q_\mu$ map the interval $[0, 1]$ inside itself if $\mu > 4$?

6.1.4. Show that
$$T_\mu(x) = \begin{cases} \frac{\mu}{2}x & \text{if } 0 \le x \le \frac{1}{2} \\ \frac{\mu}{2}(1-x) & \text{if } \frac{1}{2} \le x \le 1. \end{cases}$$

6.1.5. Show that the maximum value of $T_\mu$ is $\mu/4$.

6.1.6. For which numbers $\mu > 0$ are the mappings $C_\mu$ not differentiable?

At which points does differentiability fail?

6.1.7. Which members of the connecting family have graphs which dip below that of id on the interval $(0, \frac{1}{2})$?

Recall from Chapter 2 that an orbit $(x_0, x_1, x_2, \dots)$ of the mapping $Q_\mu$ is the solution of the difference equation

$$x_{n+1} = Q_\mu(x_n) = \mu x_n(1 - x_n).$$

with initial value $x_0$.

6.1.8.

(a) Use the 'change of variable' $x_n = \frac{1}{2} - \frac{1}{2}y_n$ in the difference equation

$$x_{n+1} = 2x_n(1 - x_n)$$

to get a difference equation for $y_n$.

(b) Solve the difference equation to get $y_n$ in terms of $y_0$.

(c) Hence show that for $n \geq 1$ the $n$th iterate of $x_0$ under the logistic mapping $Q_2$ is given by the formula

$$x_n = \frac{1}{2} - \frac{1}{2}(2x_0 - 1)^{2^n}.$$

6.1.9.

(a) Use the 'change of variable' $x_n = \sin^2(\theta_n)$ in the difference equation

$$x_{n+1} = 4x_n(1 - x_n)$$

to get

$$\sin^2(\theta_{n+1}) = \sin^2(2\theta_n).$$

(b) Solve the difference equation

$$\theta_{n+1} = 2\theta_n$$

to get $\theta_n$ in terms of $\theta_0$.

(c) Hence show that the $n$th iterate of $x_0$ under the logistic mapping $Q_4$ is given by the formula

$$x_n = \sin^2(2^n\theta_0)$$

where $\theta_0$ satisfies the equation $\sin^2(\theta_0) = x_0$.

6.1.10. Check the result of Exercise 8(c) by applying it in the cases where $x_0$ is a fixed point of $Q_2$.

6.1.11. Check that result of Exercise 9(c) by applying it in the cases where $x_0$ is a fixed point of $Q_4$.

6.1.12. Show that the following mappings are symmetric about $x = \frac{1}{2}$:

(a) $Q_\mu$ (b) $T_\mu$ (c) $C_\mu$.

6.1.13. Give an example of a mapping from $[0, 1]$ into $[0, 1]$ which is not unimodal.

6.1.14. We say that a family of unimodal mappings $\mu \mapsto f_\mu$ with $f_\mu : [0, 1] \to [0, 1]$ and $\mu \in [0, 4]$ is *replete* if

(i) $f_\mu(0) = f_\mu(1) = 0$ for each $\mu$,

(ii) $f_0$ is the zero mapping,

(iii) $f_4$ has maximum value 1.

Which of the three families of mappings defined in the text are replete?

6.1.15. Let $f : [0, 1] \to [0, 1]$ with $f(x) = \lim_{\mu \to 0} C_\mu(x)$.

(a) Use the graphs in Figure 6.1.3 to guess how the graph of $f$ looks.

(b) Check your guess in part (a) by evaluating the limit.

6.1.16. Let $g : [0, 1] \to [0, 1]$ with $g(x) = \lim_{\mu \to \infty} C_\mu(x)$.

(a) Use the graphs in Figure 6.1.3 to guess how the graph of $g$ looks.

(b) Check your guess in part (a) by evaluating the limit.

6.1.17. Let $\mu > 1$ and (temporarily) regard the logistic mapping $Q_\mu$ as having $\mathbb{R}$ as domain. Let $(x_0, x_1, x_2, \dots)$ denote the orbit of $x_0$ under $Q_\mu$.

(a) Prove that if $x_0 < 0$, then $x_n \le x_0\, \mu^n$ for each $n \in \mathbb{N}_0$,

(b) Deduce that if $x_0 < 0$, then the orbit of $x_0$ diverges to $-\infty$.

(c) If $x_0 > 1$, what is the sign of $x_1$? What is the fate of the orbit?

## 6.2 FAMILIES OF FIXED POINTS

In this section we study the changes which occur in the fixed points of a mapping $f_\mu$ as the parameter $\mu$ increases.

Plotting the graph of $f_\mu$, for a range of values of $\mu$ provides some overall perspective on these changes. For example, the number of fixed points of $f_\mu$ is the number of times the graph of $f_\mu$ crosses the graph of id. If $f_\mu$ is simple enough the results obtained in this way can be checked algebraically.

The logistic family of mappings $\mu \mapsto Q_\mu$ will be used as an illustration.

### Fixed points of $Q_\mu$

*Graphs.* Figure 6.2.1 suggests that the graph of $Q_\mu$ always intersects that of id at the origin and hence that $Q_\mu$ has 0 as a fixed point.

For large enough values of $\mu$ there is a second point of intersection and hence a second fixed point.

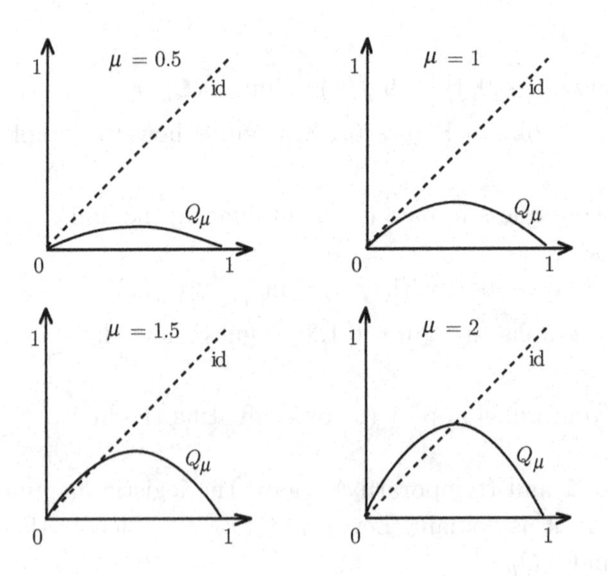


**Figure 6.2.1** One fixed point for $\mu \leq 1$ and a second fixed point for $\mu > 1$.

The graphs suggest that the change from one to two fixed points occurs when id is tangent to $Q_\mu$ at the fixed point 0. The precise value of $\mu$ where the mappings are tangent can be found by differentiating

$$Q_\mu(x) = \mu x(1-x) \tag{1}$$

to get

$$Q_\mu{}'(x) = \mu - 2\mu x. \tag{2}$$

Thus

$$Q_\mu{}'(0) = \mu. \tag{3}$$

and so the slope of $Q_\mu$ at the origin is $\mu$. It follows that

- if $\mu = 1$, the graph of $Q_\mu$ is tangent at the origin to that of id,

and near the origin

- if $\mu > 1$, the graph of $Q_\mu$ rises above that of id, but
- if $\mu < 1$ the graph of $Q_\mu$ drops below that of id.

Thus the graphs suggest, and differentiation confirms, that
*as $\mu$ increases past 1, the mapping $Q_\mu$ acquires a second fixed point.*

*Algebra.* Because the mapping $Q_\mu$ is a quadratic we can solve the equations for the fixed points and hence verify algebraically what the graphs suggest. The equation $Q_\mu(x) = x$, where $x$ is to lie in the domain $[0,1]$ of $Q_\mu$, is equivalent to $\mu x(1-x) = x$ and hence to

$$x = 0 \quad \text{or} \quad \mu(1-x) = 1.$$

The latter equation is equivalent to $1 - x = 1/\mu$ and hence has the solution

$$x = 1 - \frac{1}{\mu} = \frac{\mu - 1}{\mu}. \tag{4}$$

Thus

*if $\mu \leq 1$ there is only one fixed point $x = 0$,*

but

*if $\mu > 1$ there is a second fixed point $x = (\mu-1)/\mu$*

in the required domain $[0,1]$. ∎

## A bifurcation diagram

To see how the fixed points vary with $\mu$, we plot the graphs of each of the families of fixed points found above. Thus we plot the sets of points in the plane

$$\{(\mu, x) : x = 0 \ \ and \ \ \mu \geq 0\}$$

and

$$\{(\mu, x) : x = (\mu - 1)/\mu \ \ and \ \ \mu \geq 1\}.$$

To see the relationship between these two families of fixed points, moreover, it is convenient to plot graphs of both families on the same diagram, as in Figure 6.2.2. The resulting set of points is the *union* of the above two sets and hence may be written as

$$\{(\mu, x) : x = 0 \ \ or \ \ x = (\mu - 1)/\mu\}.$$

Note that this is the graph of a *relation* rather than a function. The graph of function may only intersect a vertical line through any point in its domain *once*. The graph of a relation is not subject to this restriction.

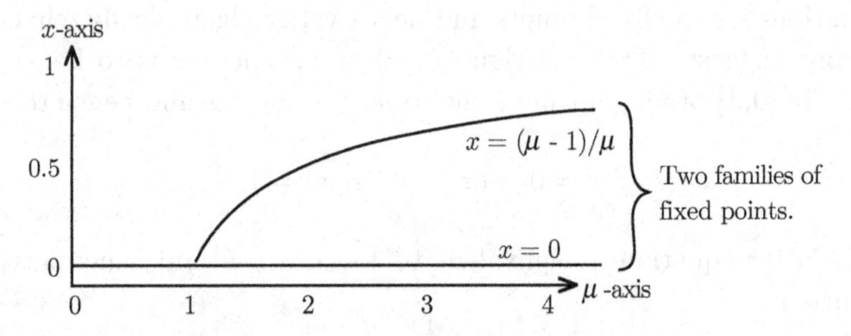


**Figure 6.2.2** Bifurcation diagram for the fixed points of the family $\mu \mapsto Q_\mu$. As $\mu$ increases past the 'bifurcation value' 1, an extra family of fixed point is 'born'.

The diagram shows that, as the parameter $\mu$ passes through the value 1, the relation splits into two branches. Because of this, the fixed points are said to *bifurcate* (at the fixed point $x = 0$ when the parameter $\mu = 1$). This is the origin of the name *bifurcation diagram* for a plot showing the periodic points versus the parameter.

In general, the *bifurcation diagram* for the fixed points of any family of mappings $\mu \mapsto f_\mu$ is the following set, plotted as a graph:

$$\{(\mu, x) : x \text{ is a fixed point of } f_\mu\}.$$

### Stability

Further detail can now be added to the bifurcation diagram to show where the fixed points are attracting and where they are repelling. Recall from Definition 5.1.5 that an attracting fixed point is one at which the derivative of the iterator has absolute value less than one, and that a repelling fixed point is one at which the absolute value of the derivative of the iterator is greater than one.

• *At the fixed point* 0, the derivative of $Q_\mu$ equals $\mu$ by (3). Hence, this fixed point is *attracting* if $0 \leq \mu < 1$ and *repelling* if $\mu > 1$. We indicate this in the modified bifurcation diagram, Figure 6.2.3, by dashing the part of the branch $x = 0$ which lies to the right of $\mu = 1$.

• *At the fixed point* $\frac{\mu-1}{\mu}$, it follows from (2) that the derivative of $Q_\mu$ is

$$Q_\mu{}'\left(\frac{\mu-1}{\mu}\right) = \mu - 2\mu\frac{\mu-1}{\mu} = 2 - \mu$$

and so this fixed point is *attracting* if $1 < \mu < 3$ and *repelling* if $\mu > 3$. We indicate this in Figure 6.2.3 by dashing the part of the branch $x = (\mu - 1)/\mu$ which lies to the right of $\mu = 3$.

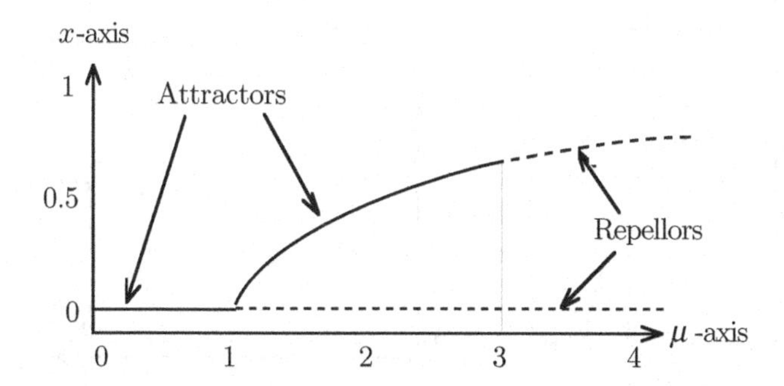


**Figure 6.2.3.** Bifurcation diagram showing the stability of the fixed points of the mappings in the logistic family $\mu \mapsto Q_\mu$.

## Snapshots

The bifurcation diagram in Figure 6.2.3 was plotted from formulae. A more direct graphical way of looking at the relationship between the iterators $Q_\mu$ and the bifurcation diagram is shown in Figure 6.2.4.

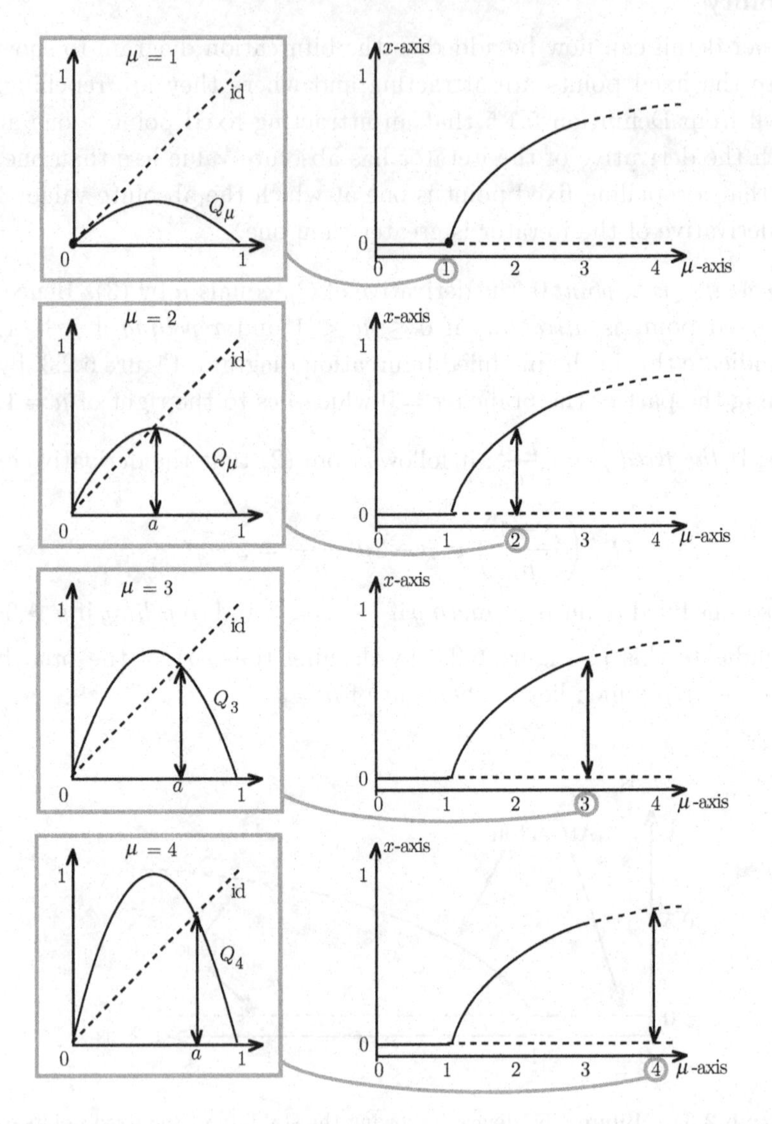


**Figure 6.2.4** Each snapshot on the left shows an iterator $Q_\mu$. Its fixed points, 0 and $a$, are indicated by the double-headed arrow. Translating the arrow across to the correct $\mu$-value gives two points on the bifurcation diagram at the right.

The snapshots of the iterators thus provide an alternative way of plotting points on the bifurcation diagram:

*for various values of $\mu$ find the fixed point a from the graph of $Q_\mu$ and then plot the points $(\mu, 0)$ and $(\mu, a)$ on the bifurcation diagram.*

The stability of the fixed points can also be determined from the snapshots, by examining the slope of $Q_\mu$ at the fixed points.

These ideas can clearly be adapted to any family of iterators.

A new type of bifurcation is introduced in Exercise 6.2.1. It is called a *saddle-node* bifurcation.

## Exercises 6.2

6.2.1. Snapshots of some iterators, in a certain family $\mu \mapsto f_\mu$, are shown in Figure 6.2.5.

(a) Use the snapshots to plot some points on the bifurcation diagram for the fixed points of this family.

(b) Join the points plotted in (a) to get the bifurcation diagram. Dash parts of the curves corresponding to repelling fixed points.

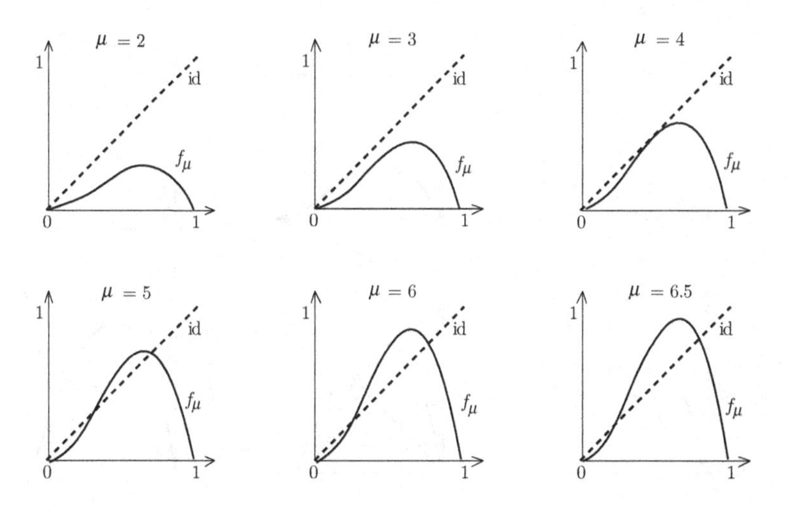


**Figure 6.2.5** Graph of $f_\mu$ for various values of $\mu$.

6.2.2. The graphs shown in Exercise 1 were plotted from the formula

$$f_\mu(x) = \mu x^2(1 - x).$$

(a) Find the derivative $f_\mu{}'(x)$ and verify that the derivative is 0 when $x = 0$ and when $x = 2/3$.

(b) Show that $f_\mu$ maps the interval $[0, 1]$ into itself if $0 \leq \mu \leq \frac{27}{4}$.

(c) Find the fixed points of $f_\mu$ algebraically and show that they are given by $x = 0$ and, for $\mu \geq 4$, by $x = (1 \pm \sqrt{1 - 4/\mu})/2$

(d) Verify that the derivative of $f_\mu$ at these fixed points is given by

$$f_\mu{}'(x) = 3 - \frac{\mu}{2} \mp \frac{\sqrt{\mu^2 - 4\mu}}{2}.$$

(e) Use the results from parts (c) and (d) to sketch a bifurcation diagram for the fixed points of the family $\mu \to f_\mu$. Compare the result with your answer to Exercise 1.

6.2.3. Snapshots of some iterators in a certain family $\mu \mapsto g_\mu$ are shown in Figure 6.2.6. Use these snapshots to plot some points on the bifurcation diagram for the fixed points of this family $\mu \mapsto g_\mu$.

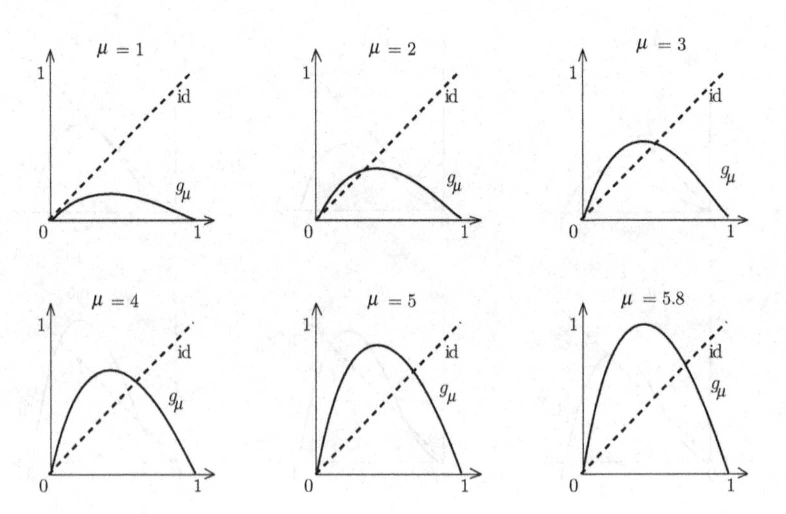


**Figure 6.2.6** Graph of $g_\mu$ for various values of $\mu$.

6.2.4. The graphs shown in Exercise 3 were plotted from the formula

$$g_\mu(x) = \mu x \left(\frac{1-x}{1+x}\right).$$

(a) Show that $g_\mu$ has its maximum value at the point $\sqrt{2}-1$; and that the maximum value of $g_\mu$ is $\mu(3-2\sqrt{2})$.

(b) Deduce that $g_\mu$ maps $[0,1]$ into itself for $0 \le \mu \le 3+2\sqrt{2}$.

(c) Show algebraically that the fixed points are $x=0$ and $x = (\mu-1)/(\mu+1)$ if $\mu \ge 1$.

(d) Find the derivative $g_\mu{}'(x)$ and hence show that $g_\mu{}'(0) = \mu$ and

$$g_\mu{}'\left(\frac{\mu-1}{\mu+1}\right) = 1 - \frac{\mu}{2} + \frac{1}{2\mu}.$$

(e) Use the results from parts (c) and (d) to sketch a bifurcation diagram for the fixed points of the family $\mu \mapsto g_\mu$. Compare the result with your answer to Exercise 3.

6.2.5.

(a) Use the graphs in Figure 6.1.2 to plot some points on the bifurcation diagram for the family of tent mappings $\mu \mapsto T_\mu$.

(b) Use the formula for $T_\mu(x)$ given in Exercise 6.1.4 to prove that $x$ is a fixed point of $T_\mu$ if and only if $x=0$

*or* $\mu = 2$ and $0 \le x \le \frac{1}{2}$,

*or* $\mu > 2$ and $x = \frac{\mu}{2+\mu}$.

Hence check your answer to part (a).

6.2.6. By referring to Figure 6.2.3, sketch the bifurcation diagram for the fixed points of each of the following families of mappings:

(a) $\mu \mapsto f_\mu$, where $f_\mu = Q_{4-\mu}$ for $0 \le \mu \le 4$,

(b) $\mu \mapsto g_\mu$, where $f_\mu = Q_{2\mu}$ for $0 \le \mu \le 2$.

6.2.7. Let $f_\mu = Q_{\mu(4-\mu)}$ for $0 \le \mu \le 4$.

(a) Explain why the bifurcation diagram for the fixed points of the family $\mu \mapsto f_\mu$ must be symmetric about $\mu = 2$.

(b) By using Figure 6.2.3, sketch the bifurcation diagram for the fixed points of the family $\mu \mapsto f_\mu$.

## 6.3 FAMILIES OF PERIOD-2 POINTS

In this section we study the changes which occur in the period-2 points of the logistic mapping $Q_\mu$ as the parameter $\mu$ varies. As in the previous section, we first use graphs to provide an overview of the changes we should expect. We then use algebra to confirm the results.

The graphical approach can be applied to any family of iterators $\mu \mapsto f_\mu$, but we can only use algebra as a check when $f_\mu(x)$ is given by a simple enough formula.

### Period-two points as fixed points

Recall that the period-2 points of $Q_\mu$ are the fixed points of $Q_\mu^2$. Some of the fixed points of $Q_\mu^2$ are also fixed points of $Q_\mu$, which we have already studied in the previous section. Hence the novelty of this section centres on *the fixed points of $Q_\mu^2$ which are not also fixed points of $Q_\mu$*. These are the points of prime period 2 for $Q_\mu$.

Graphs of $Q_\mu^2$ are plotted in Figures 6.3.1 and 6.3.2 for some typical values of the parameter $\mu$. Immediately below the graph of $Q_\mu^2$, the graph of $Q_\mu$ is also plotted so that their fixed points can be compared. Points with prime period two first appear in Figure 6.3.2, when $\mu > 3$.

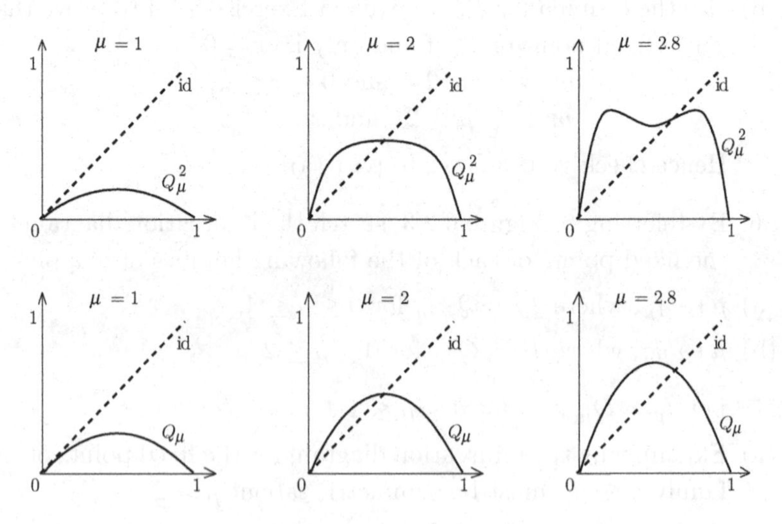


**Figure 6.3.1.** For $0 \le \mu < 3$, both $Q_\mu^2$ ( top row) and $Q_\mu$ (bottom row) have the same fixed points.

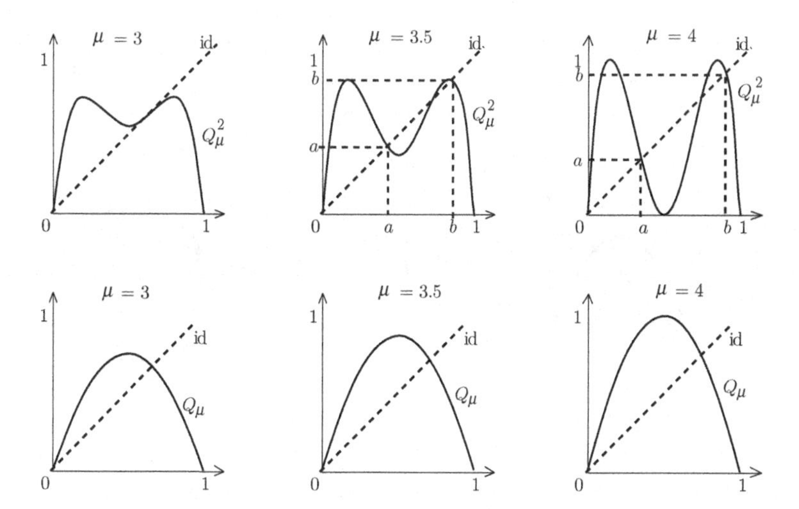


**Figure 6.3.2.** For $3 < \mu \leq 4$, the mapping $Q_\mu^2$ has two extra fixed points $a$ and $b$. They are not fixed points of $Q_\mu$, and hence are prime period 2 points.

## Bifurcation of period-2 points

When $\mu = 3$, the graphs suggest that id is tangent to $Q_\mu^2$ at a fixed point. When $\mu > 3$, the mapping $Q_\mu^2$ has two extra fixed points $a$ and $b$. Hence the fixed points of $Q_\mu^2$ bifurcate at $\mu = 3$.

The bifurcation diagram is shown in Figure 6.3.3. It consists of the two branches already obtained for the fixed points of $Q_\mu$, together with the two new branches contributed by the prime period two points of $Q_\mu$. Each point on the new branches has the form $(\mu, a)$ or $(\mu, b)$. The stability of $a$ or $b$ is determined by slope of the graph of $Q_\mu^2$ at these points.

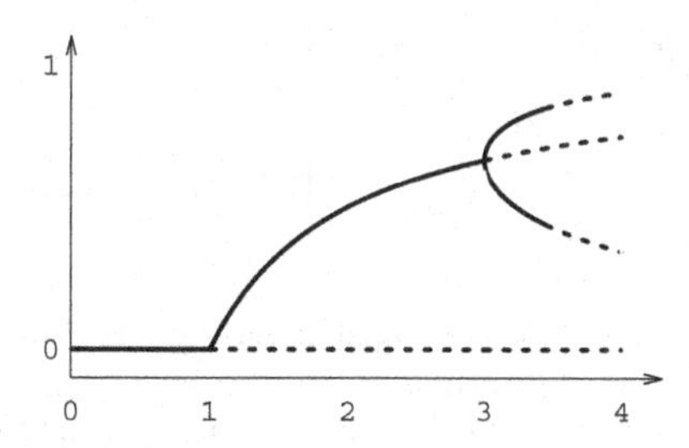


**Figure 6.3.3.** The bifurcation diagram for the fixed points of $Q_\mu^2$.

## Snapshots for bifurcation diagram

New branches in the bifurcation diagram were introduced via the fixed points of $Q_\mu^2$. The snapshots in Figure 6.3.4, however, show how these branches relate directly to the dynamics of the original iterator $Q_\mu$.

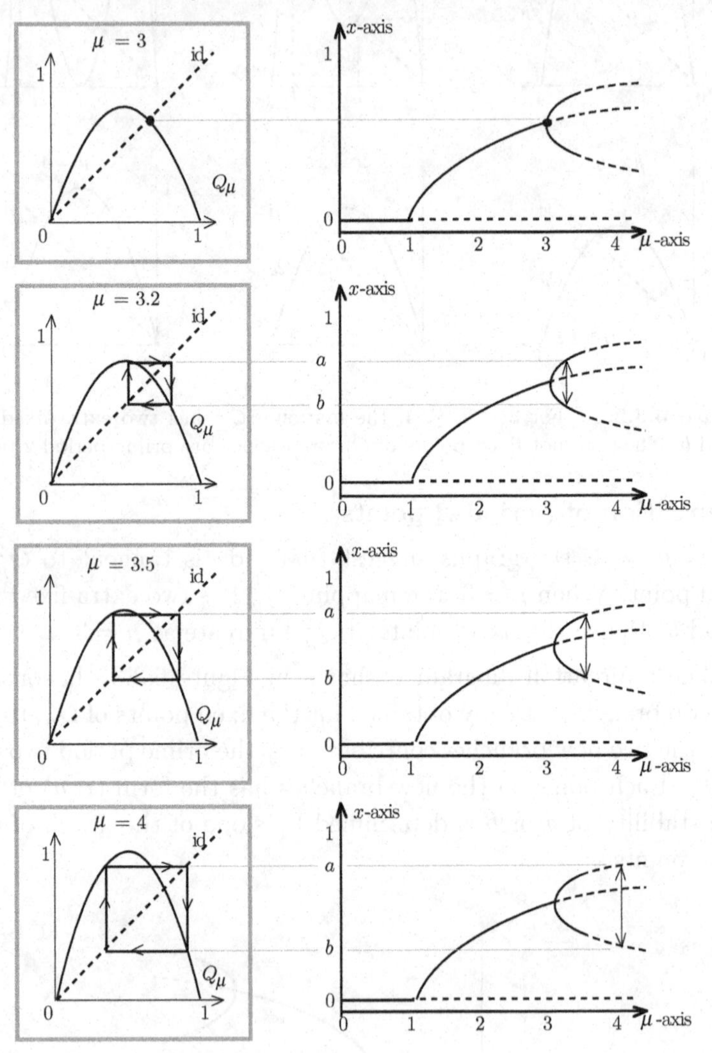


**Figure 6.3.4** For $\mu > 3$, each snapshot on the left shows a pair of fixed points $a$ and $b$ of $Q_\mu^2$. They form a prime period-2 orbit of $Q_\mu$, under which $a$ maps to $b$ and $b$ maps to $a$. Each such orbit gives a pair of points $(\mu, a)$ and $(\mu, b)$ in the bifurcation diagram on the right.

## Algebraic verification

These diagrams suggest that if $\mu > 3$ then the logistic mapping $Q_\mu^2$ has an extra pair of fixed points. Because the iterator $Q_\mu$ is a quadratic, this can be proved algebraically.

**6.3.1 Example** *Show that for $\mu > 3$, the mapping $Q_\mu^2$ has the pair of fixed points*

$$x = \frac{\mu+1}{2\mu} \pm \frac{\sqrt{(\mu+1)(\mu-3)}}{2\mu} \tag{1}$$

*(in addition to the two fixed points 0 and $(\mu-1)/\mu$ of $Q_\mu$).*

*Solution:* The fixed points of $Q_\mu^2$ are the solutions $x$ of the equation

$$(Q_\mu^2)(x) = x. \tag{2}$$

Here $Q_\mu(x) = \mu x(1-x)$ and so (2) is equivalent to each of the following equations:

$$\begin{aligned} \mu Q_\mu(x)(1 - Q_\mu(x)) &= x, \\ \mu Q_\mu(x) - \mu\,(Q_\mu(x))^2 &= x, \\ \mu^2 x(1-x) - \mu^3 x^2(1-x)^2 &= x. \end{aligned}$$

It is left as Exercise 6.3.4 to expand the left side of the last equation and hence to verify that (2) is equivalent to

$$xp(x) = 0 \tag{3}$$

where

$$p(x) = -\mu^3 x^3 + 2\mu^3 x^2 - (\mu^3 + \mu^2)x + (\mu^2 - 1). \tag{4}$$

But $Q_\mu$, and hence also $Q_\mu^2$, has the fixed points 0 and $(\mu-1)/\mu$, by Section 6.2. It follows that $(\mu-1)/\mu$ is a root of $p(x)$. Hence the Factor Theorem shows that $p(x)$ factorizes to give

$$p(x) = (x - (\mu-1)/\mu)\, q(x) \tag{5}$$

where $q(x)$ is a quadratic, whose zeroes give the extra pair of fixed points. It is left as Exercise 6.3.5 to show that

$$q(x) = -\mu^3 x^2 + (\mu^3 + \mu^2)x - (\mu^2 + \mu).$$

Hence, by Exercise 6.3.6, the quadratic equation $q(x) = 0$ has the solutions (1).

Now $\mu > 0$ and so $\mu + 1 > 0$. The solutions (1) are therefore real if and only if $\mu \geq 3$. Hence the extra pair of fixed points suddenly appears when $\mu$ increases past 3. ■

It is left as Exercise 6.3.7 to show that, at each of the prime period-2 points of $Q_\mu$, the derivative of $Q_\mu^2$ has the value

$$5 - (\mu - 1)^2.$$

Hence these two fixed points of $Q_\mu^2$ are attracting if $3 < \mu < 1 + \sqrt{6}$ and repelling if $1 + \sqrt{6} < \mu \leq 4$, in agreement with the dashing in the bifurcation diagram of Figure 6.3.3.

## Exercises 6.3

6.3.1. Let $f : [0, 1] \to [0, 1]$ be differentiable and let $a$ and $b$ be the points in a prime period-2 orbit of $f$.

(a) Use the chain rule to prove that $(f^2)'(a) = f'(b)f'(a)$.

(b) Write down a similar formula for $(f^2)'(b)$ and then deduce that

$$(f^2)'(a) = (f^2)'(b).$$

6.3.2. Suppose a differentiable mapping

$$f : [0, 1] \to [0, 1]$$

has only two fixed points, including 0. Suppose, furthermore, that $f^2$ has the graph shown in Figure 6.3.5, with fixed points $0, p, q$ and $r$.

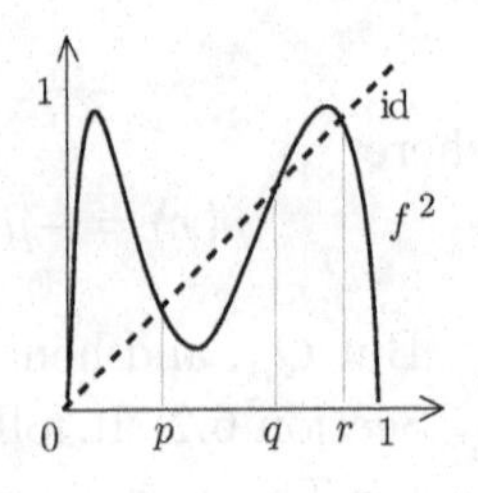

**Figure 6.3.5.**

(a) Which are the fixed points of $f$?

(b) Which are the prime period-2 points of $f$?

6.3.3. Let $g_\mu : [0,1] \to \mathbb{R}$ with

$$g_\mu(x) = \mu x \left( \frac{1-x}{1+x} \right).$$

Use the snapshots of $g_\mu^2$ shown in Figure 6.3.6 to plot some points on the bifurcation diagram for the period-2 points of $g_\mu$.

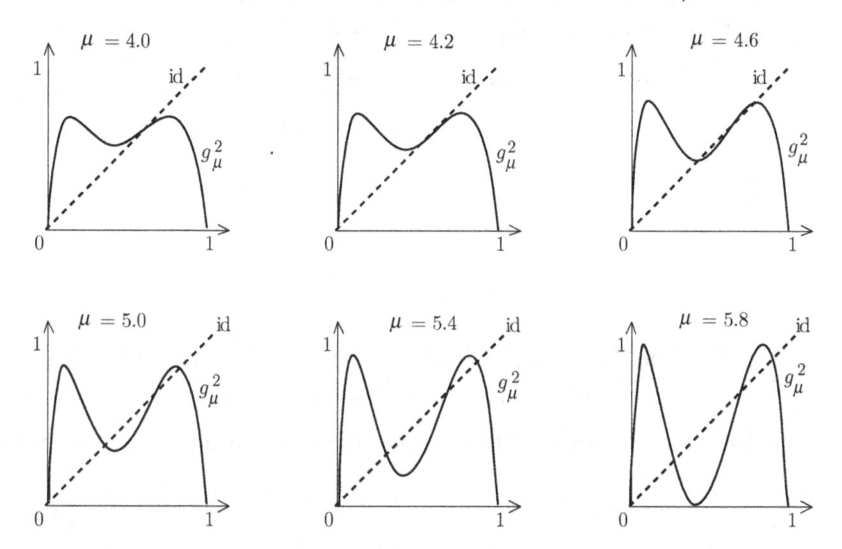


**Figure 6.3.6** Graphs of $g_\mu^2$

6.3.4. Check, as claimed in the text, that the equation

$$\mu^2 x(1-x) - \mu^3 x^2 (1-x)^2 = x$$

is equivalent to $xp(x) = 0$ where

$$p(x) = -\mu^3 x^3 + 2\mu^3 x^2 - (\mu^3 + \mu^2)x + (\mu^2 - 1).$$

6.3.5. Check, as claimed in the text, that the cubic $p(x)$ in Exercise 4 factorizes as $p(x) = (x - (\mu - 1)/\mu)\, q(x)$ where

$$q(x) = -\mu^3 x^2 + (\mu^3 + \mu^2)x - (\mu^2 + \mu).$$

[Hint. To find $q(x)$, substitute $q(x) = ax^2 + bx + c$ in the first of the above equations and then equate coefficients.]

6.3.6. Check, as claimed in the text, that the solutions of the quadratic equation $q(x) = 0$, with $q(x)$ as in Exercise 5, are given by

$$x = \frac{\mu + 1}{2\mu} \pm \frac{\sqrt{(\mu+1)(\mu-3)}}{2\mu}.$$

6.3.7. Let $a$ and $b$ be the points in a prime period-2 orbit of the logistic mapping $Q_\mu$, where $\mu > 3$.

(a) Show from the result of Exercise 1(a) that

$$(Q_\mu^2)'(a) = \mu^2(1 - 2a)(1 - 2b).$$

(b) Deduce that, at either of the prime period-2 points of $Q_\mu$, the derivative of $Q_\mu^2$ has the value $5 - (\mu - 1)^2$, as claimed in the text.

6.3.8. Can the diagram in Figure 6.3.7 be the bifurcation diagram of some iterator ? Give reasons for your answer.

[Hint: how many distinct points are there in an orbit which has prime period 2?]

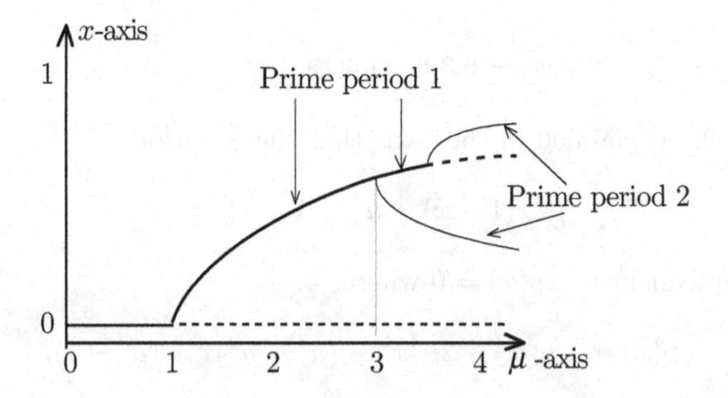


**Figure 6.3.7** Is this a possible bifurcation diagram?

## 6.4 THE BIFURCATION DIAGRAM

In this section we explore the bifurcation diagram for the families of all periodic points. We will do this with the aid of the computer in Figure 6.4.1, but in later chapters we will explain some of the mathematics behind it.

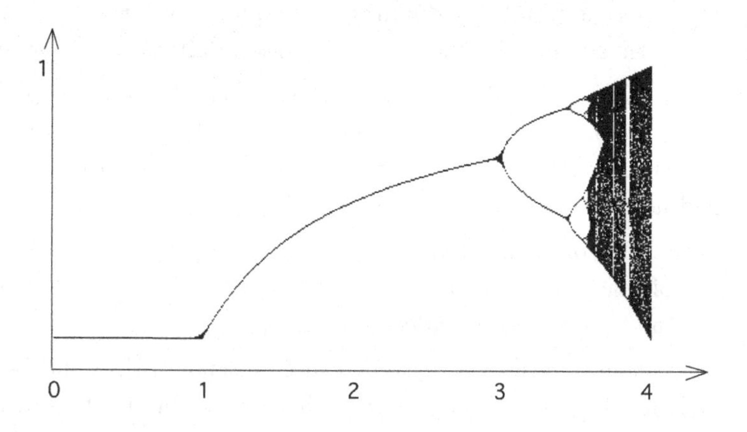


**Figure 6.4.1** The bifurcation diagram.

The bifurcation diagram in Figure 6.4.1 shows that the period-2 family is followed by the period-4 and then 8 and so on. This is known as the *period doubling cascade*. For certain parameter values all branches in the diagram simultaneously bifurcate into two branches. The first two of these *period doublings* are clearly visible in Figure 6.4.1. After the end of this period doubling process the bifurcation diagram seems to alternate between chaos and attracting periodic orbits of other periods. An attracting period-3 orbit is just visible in Figure 6.4.1. Can you see it?

---

### Additional reading for Chapter 6

The logistic family of mappings is a standard example in discrete dynamical systems, but goes by different names in different texts. It is introduced as the 'quadratic family' in Section 1.4 of [Gul] and further analysed in Section 1.5 in the context of population modelling. It is also called the 'quadratic family' in Section 1.5 of [De1], but called the 'logistic function' in Section 2.2 of [De2] where 'quadratic family' is used to refer to a different family of functions. It is discussed without any name in Sections 4.7 and 4.8 of [PMJPSY] and introduced in chapter 7 of [Ho] under the title 'logistic function'. In Sections 1.4 and 12.2 of [PJS] the term 'logistic equation' is used to refer to a slightly different family of quadratic mappings and Section 6.4 the term 'logistic equation' refers to the logistic family as we know it.

All of these texts discuss the way in which the families of fixed and period-2 points appear as attractors and then become repelling as the parameter increases. Section 1.5 of [Gul] gives a particularly detailed account of the this behaviour.

Section 1.4 of [Gul] gives definitions of several other interesting families of mappings, including the tent family (although the parameter for the tent family is scaled differently).

Bifurcation diagrams are studied in chapter 7 of [Ho] and in Section 1.6 of [Gu]. Both authors use the logistic family to illustrate some of the main ideas.

---

## References

[De1] Robert L. Devaney, *An Introduction to Chaotic Dynamical Systems: Second Edition,* Addison-Wesley, Menlo Park California, 1989.

[De2] Robert L. Devaney, *Chaos, Fractals and Dynamics, Computer Experiments in Mathematics,* Addison-Wesley, New York, 1990.

[Gul] Denny Gulick, *Encounters with Chaos,* McGraw-Hill, Inc., 1992.

[Ho] Richard Holmgren, *A First Course in Discrete Dynamical Systems* Springer-Verlag, 1994.

[PJS] Hans-Otto Peitgen, Hartmut Jürgens and Dietmar Saupe, *Chaos and Fractals, New Frontiers of Science*, Springer-Verlag, New York, 1992.

[PMJPSY] Hans-Otto Peitgen, Evan Maletsky, Hartmut Jürgens, Terry Perciante, Dietmar Saupe, and Lee Yunker, *Fractals for the Classroom: Strategic Activities Volume Two,* Springer-Verlag, 1992. (two volumes)

# 7

# GRAPHICAL COMPOSITION, WIGGLY ITERATES AND ZEROS

*In the special case where the parameter* $\mu$ *has the value 4, both the logistic mapping* $Q_\mu$ *and the tent mapping* $T_\mu$ *map the interval* $[0, 1]$ *onto itself. If* $f$ *is one of these mappings, it is easy to draw some general conclusions about the graph of* $f^n$. *These conclusions will be used to study the chaotic behaviour of* $Q_4$ *and* $T_4$ *in later chapters.*

*In Section 7.1, we explain how to compose the graphs of two mappings. In Section 7.2 we use this technique to study the graph of* $f^n$. *We find that* $f^n$ *assumes alternately the values 0 and 1 giving a graph with a wiggly appearance. Finally, in Section 7.3, we show how to locate the points at which* $f^n$ *assumes the value 0.*

## 7.1 GRAPHING THE SECOND ITERATE

This section explains how to deduce the general shape of the graph of $f^2$ from that of $f$. The technique uses the definition of $f^2$ as $f \circ f$.

### Composing graphs

The mapping $x \mapsto f^2(x)$ is the result of applying $f$ twice and hence can be written as the composite map

$$f^2 : \quad x \overset{f}{\longmapsto} f(x) \overset{f}{\longmapsto} f(f(x)).$$

Hence we need two copies of the graph of $f$:

(a) showing $x$ in the domain; and (b) showing $f(x)$ in the domain.

**7.1.1 Example** *Let $f$ be the logistic map with $\mu = 2$. Use the graphs in Figure 7.1.1 to sketch the graph of $f^2$.*

*Solution:*

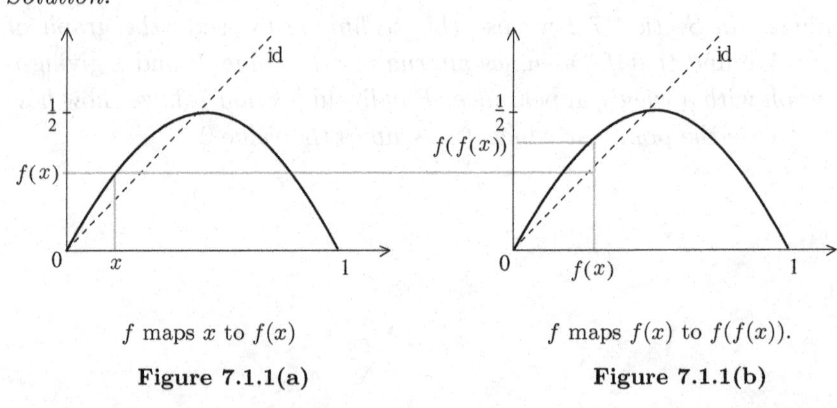


$f$ maps $x$ to $f(x)$

**Figure 7.1.1(a)**

$f$ maps $f(x)$ to $f(f(x))$.

**Figure 7.1.1(b)**

From the respective graphs it is clear that:

(a) *As $x$ increases from 0 to $\frac{1}{2}$, $f(x)$ increases from 0 to $\frac{1}{2}$;*

(b) *as $f(x)$ increases from 0 to $\frac{1}{2}$, $f^2(x)$ increases from 0 to $\frac{1}{2}$.*

Combining these two statements, after comparing the last line of statement (a) with the first line of statement (b), shows that *as $x$ increases from 0 to $\frac{1}{2}$,* *$f^2(x)$ increases from 0 to $\frac{1}{2}$.*

This leads to the sketch in Figure 7.1.2 of the part of the graph of $f^2$ which lies over the interval $[0, \frac{1}{2}]$.

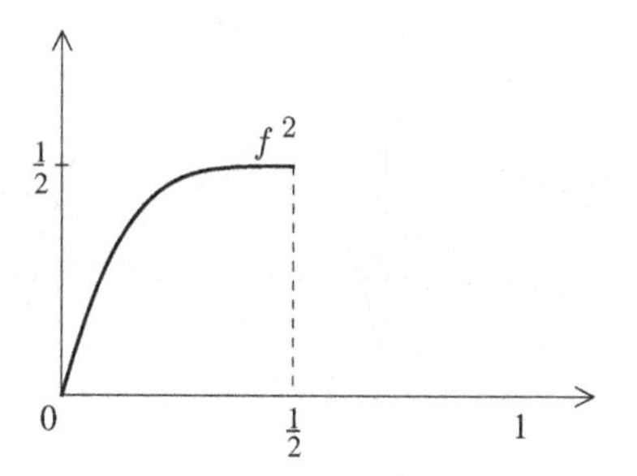


On the left of the interval, where the slope of $f$ is bigger than 1, the graph increases more rapidly than that of $f$.

**Figure 7.1.2** Graph of $f^2$ on $[0, \frac{1}{2}]$.

To complete the graph of $f^2$ we need to find the part which lies above the interval $[\frac{1}{2}, 1]$. The simplest way to do this is to use the fact that $f^2$ must share the symmetry of $f$ about $x = \frac{1}{2}$. Hence we get the graph of $f^2$ shown in Figure 7.1.3.

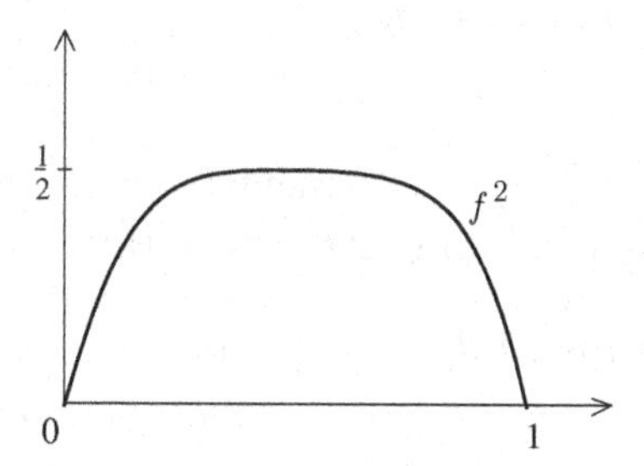


**Figure 7.1.3.** Graph of $f^2$.

Alternatively, we can use an argument similar to that used above to show that

*as $x$ increases from $\frac{1}{2}$ to 1,* *$f^2(x)$ decreases from $\frac{1}{2}$ to 0.*

This provides an alternative way of completing the graph of $f^2$, which can be used even when the graph of $f$ is not symmetrical. ■

The next example is interesting from the point of view of our long-term goals in that an extra hump appears under composition.

**7.1.2 Example** *Let $f$ be the logistic map $Q_\mu$ with $\mu = 4$. Use the graphs in Figure 7.1.4 to sketch the general shape of $f^2$.*

*Solution:*

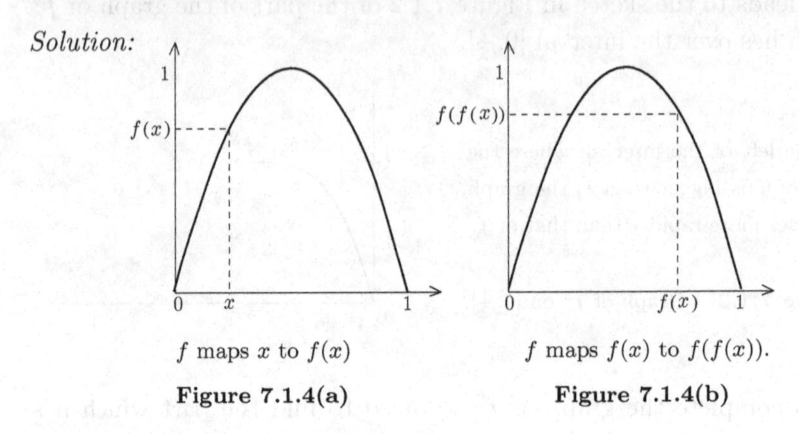


$f$ maps $x$ to $f(x)$

**Figure 7.1.4(a)**

$f$ maps $f(x)$ to $f(f(x))$.

**Figure 7.1.4(b)**

It is clear from the respective graphs that:

(a) *As $x$ increases from* 0 *to* $\frac{1}{2}$, $f(x)$ *increases from 0 to 1,*

(b) *as $f(x)$ increases from 0 to 1, $f(f(x))$ increases from 0 to 1 and then decreases back to 0.*

Comparing the last line of statement (a) with the first line of statement (b) and then combining these two statements gives:

*As $x$ increases from 0 to $\frac{1}{2}$, $f(f(x))$ increases from 0 to 1 and then decreases back to 0.*

Thus the part of the graph of $f^2$ lying directly above the interval $[0, \frac{1}{2}]$ consists of a single hump, as shown in Figure 7.1.5.

The mapping $f$ increases more quickly than id on the left part of the interval $[0, \frac{1}{2}]$. Hence the point at which $f^2$ assumes its maximum value is a little to the left of centre.

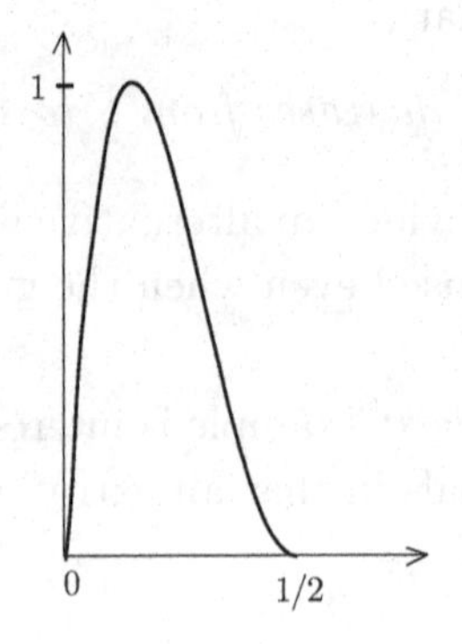


**Figure 7.1.5** Graph of $f^2$ on $[0, \frac{1}{2}]$.

To complete the graph of $f^2$ we may again use the symmetry about the line $x = \frac{1}{2}$, which $f^2$ shares with $f$, to get the complete graph of $f^2$ shown in Figure 7.1.6, with two humps.

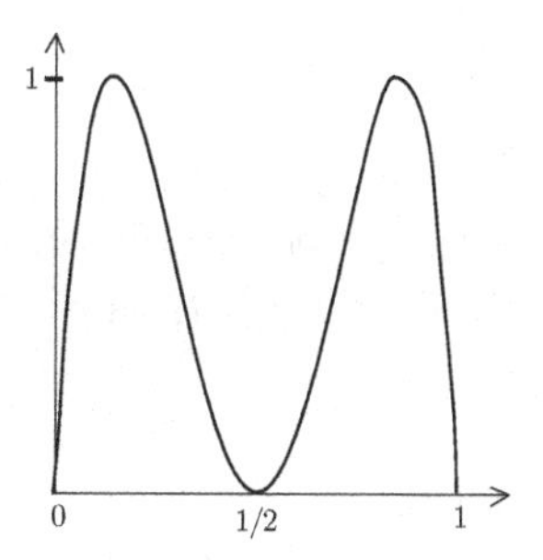


**Figure 7.1.6** Graph of $f^2$. ■

Writing $f^3 = f^2 \circ f$ and then using the argument in Example 7.1.2, we can show that

*the graph of* $f^3$ *has twice as many humps as the graph of* $f^2$.

In a similar way we can show that

*the graph of* $f^4$ *has twice the number of humps as the graph of* $f^3$

— and so on. The general result for $f^n$ which these examples suggest will be stated carefully in the next section and then proved by induction.

## Adding rigour

The technique of graphical composition will be the basis for our study of chaos. Hence it is desirable to indicate how it can be rigorously justified, independently of the graphs. Recall from calculus that

> A mapping $f$ is said to be *strictly increasing* on a set $S$ in its domain if for every $x, y \in S$
>
> $$x < y \quad \Rightarrow \quad f(x) < f(y)$$

A similar definition holds for *strictly decreasing*.

A mapping is said to be *strictly monotonic* on a set if either it is strictly increasing or strictly decreasing on the set. Definitions of *increasing, decreasing and monotonic* are obtained by using weak inequalities. It is left as Exercise 7.1.4 for you to prove the following lemma from these definitions.

**7.1.3 Lemma** *If $f$ is strictly monotonic on a set $S$ in its domain and if $g$ is strictly monotonic on $f(S)$, then $g \circ f$ is strictly monotonic on $S$.*

The technique of graphical composition can be viewed as an application of this lemma, in which we split the domain into two intervals $[0, \frac{1}{2}]$ and $[\frac{1}{2}, 1]$, on each of which $f$ is strictly monotonic.

### Using calculus

An alternative way to obtain the general shape of the graph of $f^2$, when $f$ is differentiable, is to use calculus. To differentiate $f^2$ we use the chain rule.

The chain rule, for differentiating the composite function $f \circ g$, says that if $g$ is differentiable at $x$ and $f$ is differentiable at $g(x)$ then

$$(f \circ g)'(x) = f'(g(x))g'(x).$$

In particular,

$$(f \circ f)'(x) = f'(f(x))f'(x)$$

provided that $f$ is differentiable at both $x$ and $f(x)$.

We apply this formula to the logistic map.

**7.1.4 Example** *Sketch the graphs of some typical members of the logistic family of mappings $\mu \mapsto Q_\mu^2$ using calculus.*

*Solution:* For computational convenience write $f$ for $Q_\mu$ so that

$$f(x) = \mu x(1-x).$$

To calculate the derivative of $f^2$ note that $f^2 = f \circ f$. Hence, by the chain rule,

$$\begin{aligned}(f^2)'(x) &= f'(f(x))f'(x) \\ &= \mu(1-2f(x))\mu(1-2x) \\ &= \mu^2(1-2\mu x+2\mu x^2)(1-2x).\end{aligned}$$

Thus the derivative is zero when

$$x = \frac{1}{2} \quad \text{or} \quad x = \frac{1}{2} \pm \frac{1}{2}\sqrt{(\mu-2)/\mu}$$

If $\mu > 2$ then $(\mu-2)/\mu$ is positive. Hence the square root is real and so the graph of $Q_\mu^2$ has two humps.

If $\mu \leq 2$, however, the graph has only one hump.

Figure 7.1.7 shows the graph of $Q_\mu^2$ in some typical cases.

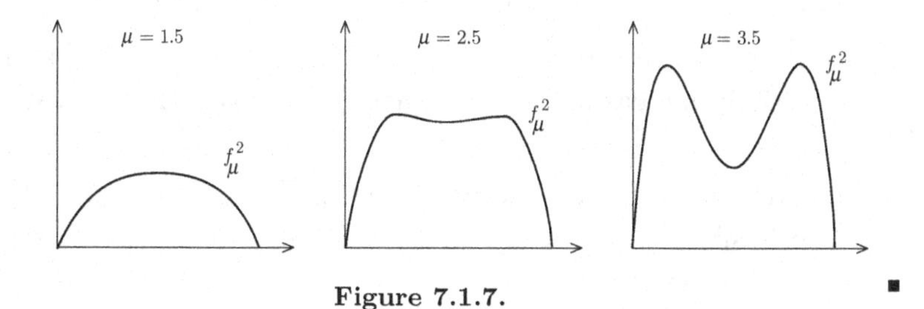


**Figure 7.1.7.** ■

## Exercises 7.1

*In all the exercises which ask for the graph of a composite mapping, follow the working of Examples 7.1.1 and 7.1.2. This involves not only sketching graphs of the component functions, but also writing sentences.*

7.1.1. Let $f$ be the logistic mapping with $\mu = 3$.

(a) Sketch two copies of the graph of $f$ next to one another. Show the maximum value of $f$ on the vertical axis.

(b) Now use graphical composition to sketch the graph of $f^2$.

7.1.2. Let $f$ be the logistic mapping with $\mu = 2$. The starting point for this exercise is the graph of $f^2$ from Figure 7.1.3.

(a) Use $f^3 = f^2 \circ f$ to help you sketch the graph of $f^3$.

(b) Use $f^4 = f^3 \circ f$ to help you sketch the graph of $f^4$.

(c) Guess the general shape of the graph of $f^n$ for each positive integer $n$.

(d) Draw a cobweb diagram for $f$ starting from various initial values. Compare the results with your answer to part (c).

7.1.3. Let $f$ be the logistic mapping with $\mu = 4$. The starting point for this exercise is the graph of $f^2$ from Figure 7.1.6.

(a) Use $f^3 = f^2 \circ f$ to help you sketch the graph of $f^3$.

(b) Use $f^4 = f^3 \circ f$ to help you sketch the graph of $f^4$.

(c) Guess the general shape of the graph of $f^n$ for each positive integer $n$.

7.1.4. Let $f$ and $g$ be mappings from $[0, 1]$ into itself.

(a) Prove that if $f$ is strictly increasing on a set $S \subseteq [0, 1]$ and $g$ is strictly increasing on $f(S)$ then $g \circ f$ is strictly increasing on $S$.

(b) Prove that if $f$ is strictly increasing on the set $S \subseteq [0, 1]$ and $g$ is strictly decreasing on the set $f(S)$ then $g \circ f$ is strictly decreasing on $S$.

(c) What remains to be done in order to complete the proof of Lemma 7.1.3 ?

7.1.5. Let $f : [0, 1] \to [0, 1]$ be differentiable and let $x \in [0, 1]$. Use the chain rule to prove that:

(a) if $f'(x) = 0$, then $(f^2)'(x) = 0$;

(b) if $f'(x) = 0$, then $(f^n)'(x) = 0$ for all $n \in \mathbb{N}$;

(c) if $(f^m)'(x) = 0$, then $(f^n)'(x) = 0$ for all $n, m \in \mathbb{N}$ with $n \geq m$.

Comment on the significance of these results for the graphs of successive iterates of a mapping.

## 7.2 ONE-HUMP MAPPINGS

In this section we discuss the graphs of higher iterates of the logistic map $Q_4$ and the tent map $T_4$. Relevant graphs are shown below.

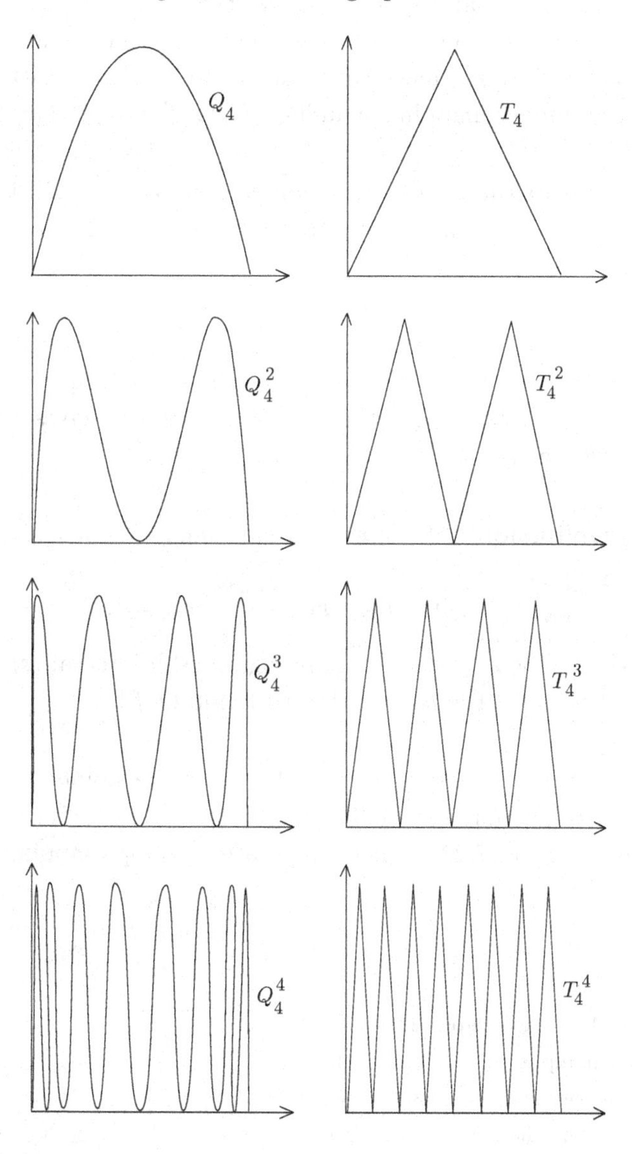


Logistic map and iterates. Tent map and iterates.

**Figure 7.2.1**

The graphs in Figure 7.2.1 suggest that there is an overall pattern for the iterates of the logistic and the tent mappings. To describe the pattern precisely we introduce some terminology.

Both the logistic and tent mappings themselves are continuous and strictly increase from 0 to a maximum value of 1 and then strictly decrease back to 0. Hence we make the following definition, in which $f$ is a continuous mapping of an interval $[a, c]$ into $[0, 1]$ and $a < b < c$ :

**7.2.1 Definition** $f$ is a *one-hump mapping* on $[a, c]$ if it is continuous and it strictly increases from $f(a) = 0$ to $f(b) = 1$, and then strictly decreases to $f(c) = 0$. ■

Thus both the logistic map and the tent map are *one-hump mappings* on the interval $[0, 1]$. The iterates of these two maps, however, have more than one hump and the following definition gives a precise way of describing this.

**7.2.2 Definition** $f^n$ is an *$m$-hump mapping* on $[0, 1]$ if there are $m + 1$ points,

$$0 = x_0 < x_1 < \cdots < x_m = 1,$$

such that $f^n$ is a one-hump mapping on each interval $[x_{i-1}, x_i]$. This interval is called the *base* of the $i$th hump of $f^n$. ■

Note that $f^n$ assumes the value 0 at the points $x_0, x_1, \ldots, x_m$, which are therefore called the *zeroes* of $f^n$. The definitions are illustrated in Figure 7.2.2, which shows a two-hump mapping.

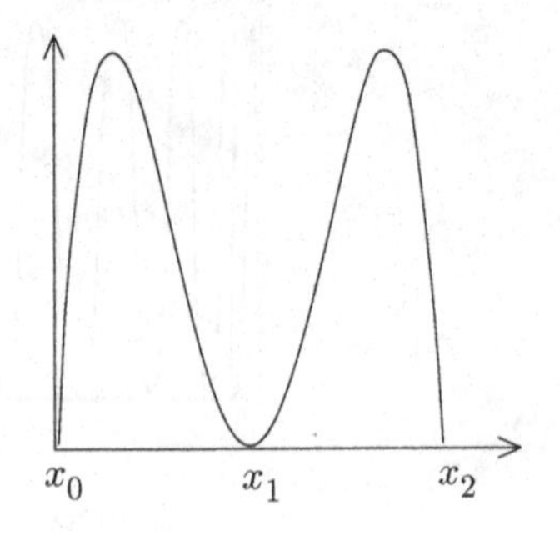


**Figure 7.2.2.**
Here $f^2$ has two humps. The base of the first hump is $[x_0, x_1]$ and the base of the second hump is $[x_1, x_2]$. The zeroes of $f^2$ are the numbers $x_0, x_1, x_2$.

The graphs shown in Figure 7.2.1 suggest that there is an overall pattern for the iterates of the logistic and the tent mappings.

- At each iteration the number of humps doubles. Hence the $n$th iterate of the logistic or the tent mapping consists of $2^{n-1}$ humps.
- The largest base length of the humps gets smaller and smaller and tends to zero as $n$ tends to infinity.

Later we prove that the logistic and the tent mappings have these two properties. First, however, we introduce some terminology that will help us to explore some consequences of the two properties.

**7.2.3 Definition** We call $f$ a mapping with *wiggly iterates* if

(i) $f^n$ is a $2^{n-1}$-hump mapping, for each $n \geq 1$, and

(ii) the length of the largest base of the humps of $f^n$ tends to 0 as $n$ tends to infinity. ■

**7.2.4 Lemma (Spike Lemma)** *Let $f : [0,1] \to [0,1]$ have wiggly iterates. For every interval $I \subseteq [0,1]$ there is a hump of some $f^n$ whose base is contained in I.*

*Proof:* Since $f$ has wiggly iterates, the length of the largest base of the humps of $f^n$ tends to 0 as $n$ tends to infinity. Hence we can choose $n$ such that the distance between two consecutive zeroes of $f^n$ is less than half the length of $I$.

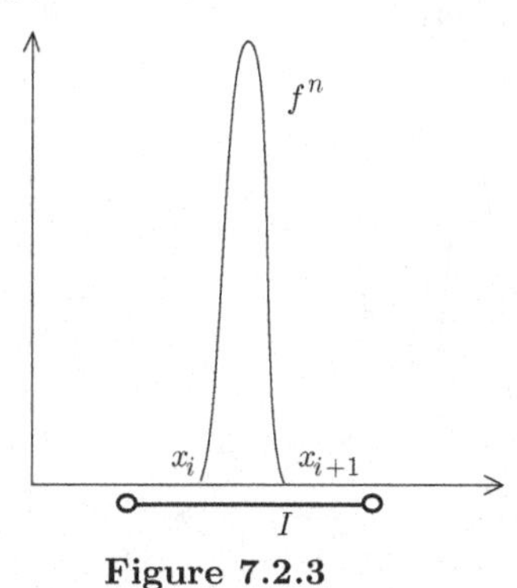

**Figure 7.2.3**

Choose the smallest zero $x_i$ which lies in $I$. It will then lie in the first half of $I$. The next zero $x_{i+1}$ of $f^n$ will also lie in $I$. Hence $[x_i, x_{i+1}] \subseteq I$. See Figure 7.2.3. ■

The Spike Lemma will later play a key role in our discussion of chaotic behaviour. We now turn, however, to the task of proving what the graphs in Figure 7.2.3 suggest: both the logisitic map and the tent map (with $\mu = 4$) have wiggly iterates. We need the following lemma, whose proof uses the technique of graphical composition introduced in the previous section.

**7.2.5 Lemma** *If $f$ is a one-hump mapping on* $[0, 1]$ *then $f^n$ is a* $2^{n-1}$ *hump mapping on* $[0, 1]$.

*Proof:* We shall use induction to prove, for all $n \in \mathbb{N}$, $f^n$ has $2^{n-1}$ humps. The claim

$$\boxed{f^n \text{ has } 2^{n-1} \text{ humps}}$$

is true when $n = 1$ as $f$ is a one-hump mapping.

Now let $n \geq 1$ be an integer for which the claim is true.

We wish to deduce the claim holds for $n + 1$.

Let $[x_i, x_{i+1}]$ be the base of a hump of $f^n$.
Let $x'$ be the point in $[x_i, x_{i+1}]$ at which $f^n$ assumes the value 1.
Composing $f$ with $f^n$ from $x_i$ to $x_{i+1}$ with the aid of graphs as in Section 7.1, we see that:

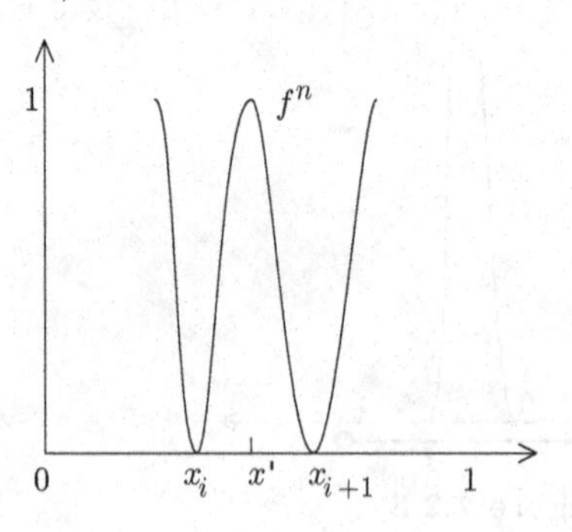


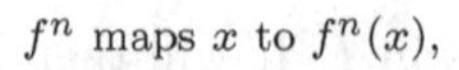
$f^n$ maps $x$ to $f^n(x)$,

**Figure 7.2.4(a)**

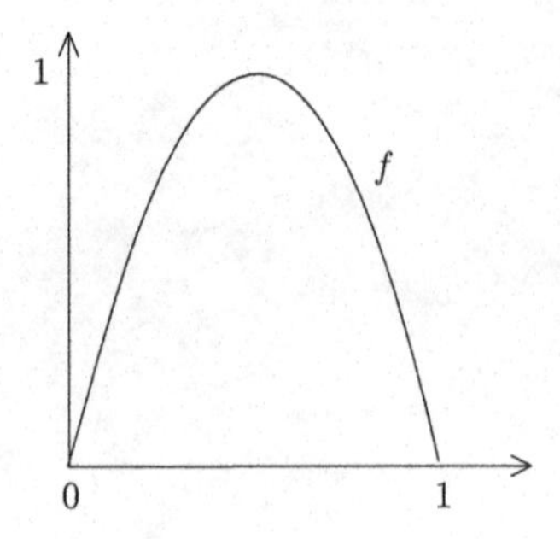


$f$ maps $f^n(x)$ to $f(f^n(x))$.

**Figure 7.2.4(b)**

From the graphs it is clear that:

(a) *as $x$ increases from $x_i$ to$x'$,*
*$f^n(x)$ increases from 0 to 1.*

(b) *as $f^n(x)$ increases from 0 to 1,*
*$f(f^n(x))$ increases from 0 to 1*
*and then decreases from 1 to 0.*

Combining these two statements, after comparing the last line of statement (a) with the first line of statement (b), shows that

*As $x$ increases from $x_i$ to $x'$,* *$f(f^n(x))$ increases from 0 to 1,*
*then decreases from 1 to 0.*

This shows that

$$f^{n+1}(x) \text{ has one hump on } [x_i, x'].$$

Similarly we see that

$$f^{n+1}(x) \text{ has one hump on } [x', x_{i+1}].$$

Thus $f^{n+1}$ has twice as many humps as $f^n$. ∎

We now use the above lemma to help us prove that the logistic and the tent mappings have wiggly iterates.

**7.2.6 Example** *Show that both the logistic mapping $Q_4$ and the tent mapping $T_4$ have wiggly iterates.*

*Solution:* Both mappings are one-hump mappings. Hence, by Lemma 7.2.5, their $n$th iterates are $2^{n-1}$-hump mappings.

To have wiggly iterates, the mappings must have the additional property that the largest base of a hump of the $n$th iterate tends to 0 as $n$ tends to infinity. This part of the proof is set as Exercise 7.2.6 for the logistic mapping and as Exercise 7.4.4 for the tent mapping.

Thus the logistic and the tent mappings have wiggly iterates. ∎

## Exercises 7.2

7.2.1. Look at the graphs of the tent mapping $T_4$ and its iterates in Figure 7.2.1 and then answer the following questions, where $\mu = 4$.

(a) Which numbers are the zeroes of the mapping $T_\mu^2$?

(b) How many humps does $T_\mu^2$ have? Which intervals form the bases of the humps?

7.2.2. Repeat Exercise 1 but with $T_\mu^3$ in place of $T_\mu^2$.

7.2.3. ('Once a zero, always a zero'.)

Let $f : [0,1] \to [0,1]$ and assume that 0 is a fixed point of $f$.

Prove, for each integer $n \geq 0$, that if $z$ is a zero of $f^n$ then $z$ is also a zero of $f^{n+1}$.

7.2.4. ('Maxima go to zeroes'.)

Let $f : [0,1] \to [0,1]$ be a one-hump mapping. Show that if $f^n$ has a maximum value at $x \in [0,1]$, then $f^{n+1}(x) = 0$.

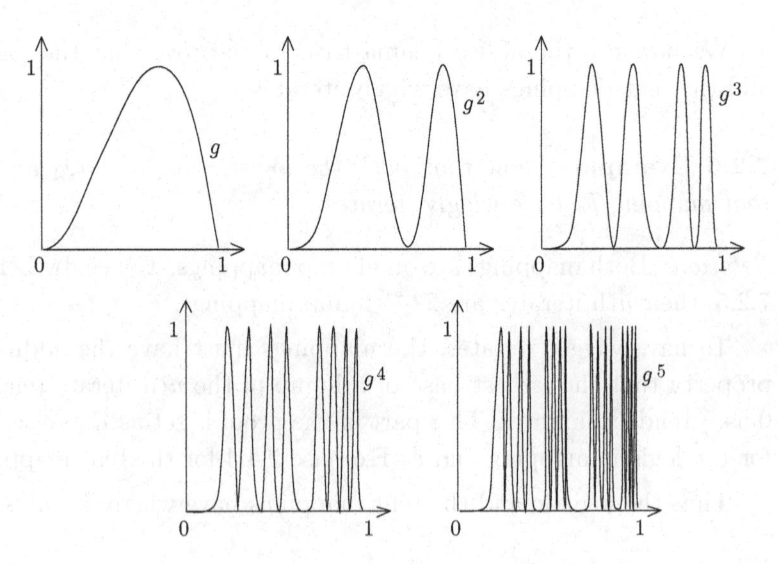


**Figure 7.2.6** Iterates of the mapping $g$ in Exercise 5.

7.2.5. The graphs of $g$, $g^2$, $g^3$, $g^4$ and $g^5$ are shown in Figure 7.2.6, where $g : [0,1] \to [0,1]$ with

$$g(x) = (27/4)x^2(1-x).$$

(a) Is $g$ a one-hump mapping ?

(b) For each of the iterates of $g$, state the number of humps shown in its graph.

(c) Is the $n$th iterate of $g$ a $2^{n-1}$-hump mapping? Give reasons for your answer.

(d) For each of the iterates of $g$ shown in Figure 7.2.6, indicate which hump has the largest base length.

(e) Suppose that the pattern suggested by the graphs continues indefinitely. Does the largest baselength of the humps of $g^n$ tend to 0 as $n \to \infty$?

(f) Does $g$ have wiggly iterates? Give reasons for your answer.

7.2.6. ('Zeroes of the Logistic Mapping.')

Let $f : [0,1] \to [0,1]$ be the logistic map $Q_4$.

(a) Recall that $f^n(x_0) = x_n$, where $x_n$ is the $n$th iterate of $x_0$ under $f$. Deduce from the result of Exercise 6.1.9 that

$$f^n(x_0) = \sin^2(2^n\theta_0) \qquad \text{where } x_0 = \sin^2(\theta_0).$$

(b) Hence show that the zeroes of $f^n$ are the numbers

$$z_k = \sin^2\left(\frac{\pi k}{2^n}\right) \qquad \text{where } k = 0, 1, 2, \ldots, 2^{n-1}.$$

(c) Deduce from the result of part (b) that the largest base of a hump of $f^n \to 0$ as $n \to \infty$. Of which result stated in the text have you now completed the proof?

7.2.7.

(a) What is the least number of points of period $n$ that a one-hump mapping can have?

(b) What is the least number of points of prime period $n$ that the logistic mapping have in each of the following cases?

(i) $n$ is a prime number; (ii) $n$ is $2^k$ where $k \in \mathbb{N}$.

## 7.3 DENSE SETS OF ZEROES

Figure 7.2.1 suggests a simple pattern formed by zeroes of successive tent mappings: To derive the zeroes of $T^{n+1}$ from those of $T^n$, add the points which are midway between consecutive zeroes of $T^n$. The resulting sets of zeroes shown in Figure 7.3.1 suggest that, if the process of adding zeroes to the diagram were continued indefinitely, the zeroes would cover the interval $[0, 1]$ *densely*, i.e., every bit of the interval $[0, 1]$ would eventually contain a zero of some $T^n$.

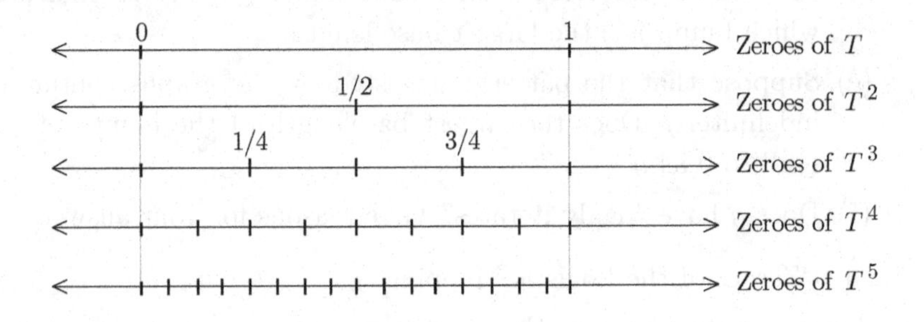


**Fig 7.3.1** Zeroes of some iterates of the tent mapping.

Before developing these ideas further, we must define denseness.

**7.3.1 Definition** Let $D \subseteq [0, 1]$. The set $D$ is said to be *dense* in $[0, 1]$ if the following condition holds:

For each interval[1] $I \subseteq [0, 1]$, there is a point of $D$ which is in $I$. ∎

Thus, no matter how short we make the interval $I$ and no matter where it is placed within the interval $[0, 1]$, it must contain at least one point of the set $D$, as illustrated in Figure 7.3.2.

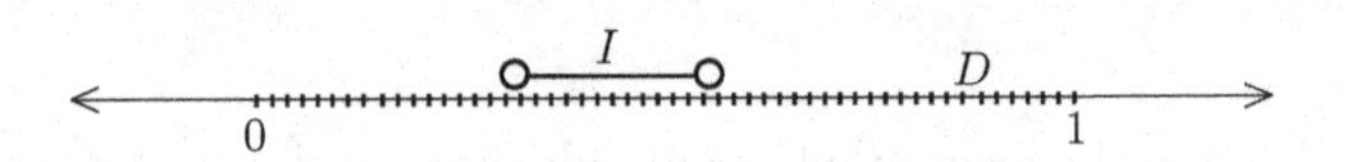


**Fig 7.3.2** Every $I$ contains a point of $D$.

[1]Strictly speaking we should insist that $I$ be *non-degenerate* to rule out the trivial cases of an empty interval or an interval of the form $\{x\} = [x, x]$. Since this becomes rather tedious, we tacitly assume that $I$ is a 'sensible' interval.

Recall from Section 7.2 that a one-hump mapping has *wiggly iterates* if the maximum distance between the pairs of consecutive zeroes of $f^n$ approaches 0 as $n$ approaches infinity.

The following lemma expresses this idea in terms of denseness. First, however, recall that to prove a statement of the form *p if and only if q* you must prove (a) *if p then q* and (b) the converse *if q then p*.

**7.3.2 Lemma (Wiggly $\Leftrightarrow$ dense sets of zeroes)**
*Let $f : [0,1] \to [0,1]$ be a one-hump mapping. The mapping $f$ has wiggly iterates if and only if the set $S$ of all zeroes of all iterates of $f$ is dense in $[0,1]$.*

*Proof:* First assume that $f$ has wiggly iterates.

To prove that $S$ is dense in $[0,1]$.

Let $I \subseteq [0,1]$ be an interval. Suppose that the length of the interval is $\epsilon > 0$. Now the maximum distance between any consecutive pair of zeroes of $f^n$ can be made less that $\epsilon/2$ by choosing $n$ sufficiently large, since the iterates are wiggly. The interval $I$ must therefore contain one of these zeroes. Hence the zeroes are dense in $[0,1]$.

To prove the converse, assume that the set $S$ is dense in $[0,1]$.

To prove that $f$ has wiggly iterates

Let $\epsilon > 0$. Let $[0,1]$ be covered by a finite collection of intervals, each with length at most $\epsilon/2$. By denseness, each of these intervals contains a zero of some $f^n$. Hence there is a finite sequence of points $z_0, z_1, z_2, \ldots, z_m$ — one point in each interval — and each $z_i$ is a zero of say $f^{n_i}$. Put $\nu = \max\{n_i\}$ so that $f^n$ has every $z_i$ as a zero for all $n > \nu$. Thus there is a $\nu$ such that for all $n > \nu$ the maximum distance between consecutive zeroes of $f^n$ is less than $\epsilon$. This means that the maximum distance approaches 0 as $n$ approaches infinity. Hence $f$ has wiggly iterates. ∎

## Exercises 7.3

7.3.1. Let $D$ be a subset of the interval $[0, 1]$.

(a) Define what it means for the set $D$ to be dense in the interval $[0, 1]$ using quantifier symbols.

(b) Deduce what it means for the set $D$ not to be dense in the interval $[0, 1]$.

7.3.2. Show that each of the following sets is not dense in $[0, 1]$. In each case draw a diagram.

(a) A subset of $[0, 1]$ with just one point $x_1$.

(b) A subset of $[0, 1]$ with just a finite number of points, say

$$x_1 < x_2 < x_3 < \cdots < x_n.$$

7.3.3. Read the footnote after the Definition 7.3.1. Suppose we tried to define a dense set $D$ by insisting that:

For each *non-empty* interval $I \subseteq [0, 1]$, there is a point of $D$ which is in $I$.

How would this change the meaning of the definition?

## 7.4 ZEROES UNDER BACKWARD ITERATION

In Section 7.2 we showed how zeroes of higher iterates of one-hump mappings form the endpoints of the base of the humps.

In this section we explain how these zeroes can be found by iterating backwards on the graph of the mapping itself.

The following lemma shows how the zeroes of the $(n+1)$st iterate of any mapping can be determined from the zeroes of the $n$th iterate.

**7.4.1 Lemma** *Let $f$ be a mapping and let $n \in \mathbb{N}_0$. Then $z_{n+1}$ is a zero of $f^{n+1}$ if and only if*

$$f(z_{n+1}) = z_n \tag{1}$$

*where $z_n$ is a zero of $f^n$.*

*Proof:* Let $z_{n+1}$ be a zero of $f^{n+1}$; that is, $f^{n+1}(z_{n+1}) = 0$ and so $f^n(f(z_{n+1})) = 0$. Hence $f(z_{n+1})$ is a zero of $f^n$.
Proving the converse is Exercise 7.4.1. ■

### Backward iteration

If $z_0, z_1, z_2, \ldots$ is any sequence which satisfies condition (1), then the following mapping diagram is valid:

$$z_0 \overset{f}{\longleftarrow} z_1 \overset{f}{\longleftarrow} z_2 \overset{f}{\longleftarrow} z_3 \ \cdots \ z_n \overset{f}{\longleftarrow} z_{n+1} \ \cdots .$$

Hence (1) is said to describe a *backward* iteration procedure.

A complication which arises with backward iteration is that the mapping $f$ is not usually one-to-one and this permits the equation (1) to have more that one solution for $z_{n+1}$.

For example, the symmetric one-hump mapping $f$ shown in Figure 7.4.1 is two-to-one on the set $[0,1] \setminus \{\frac{1}{2}\}$.

Hence, for a zero $z_n \neq \frac{1}{2}$ there are two zeroes satisfying equation (1):

*if $z_{n+1}$ is a solution then so is $1 - z_{n+1}$, since $f$ is symmetric.*

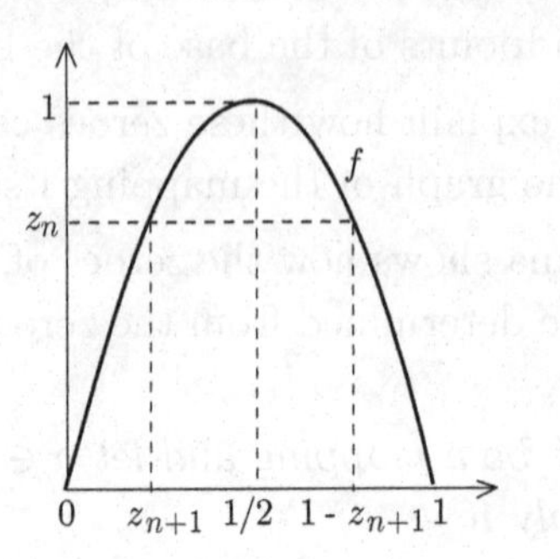


**Figure 7.4.1.** Each zero $\neq 1$ of $f^n$ leads to two zeroes of $f^{n+1}$.

To locate zeroes of $f^2$ on the graph of $f$ using Lemma 7.4.1 we solve

$$f(z_2) = 0 \text{ and } f(z_2) = 1$$

where 0 and 1 are the zeroes of $f$. Zeroes of $f$ are zeroes of $f^2$ and they are obtained by solving $f(z_2) = 0$. *Thus we first plot the zeroes 0 and 1 of $f$ on the horizontal axis.* The following steps locate the zero obtained by $f(z_2) = 1$:

1. *leave the graph of $f$ vertically to locate on* id *the point* $(1, 1)$*;*
2. *leave the graph of* id *horizontally to locate on $f$ the point* $(\frac{1}{2}, 1)$*;*
3. *plot the zero $\frac{1}{2}$ of $f^2$ on the horizontal axis.*

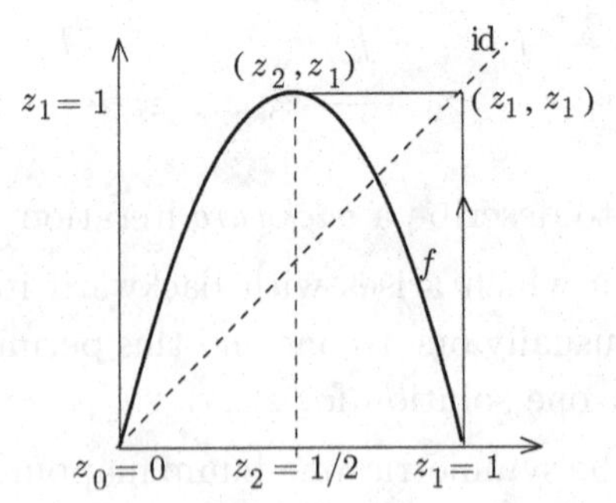


**Figure 7.4.2.** Locating zeroes of $f^2$.

Repeating the above three steps locates zeroes of all iterates. Hence we can give a general procedure to locate the zeroes of $f^n$.

**Backward iteration:** Locating zeroes.

**Step 0.** *Plot the zeroes of $f$ on the horizontal axis.*

Repeat the following steps starting with $n = 1$:

**Step 1.** *Leave the graph of $f$ vertically*
to locate on the graph of id the points $(z_n, z_n)$.

**Step 2.** *Leave the graph of* id *horizontally*
to locate on the graph of $f$ the points $(z_{n+1}, z_n)$.

**Step 3.** *Plot the zeroes $z_{n+1}$ of $f^{n+1}$ on the horizontal axis.*

In Figure 7.4.3 we use the above method to locate zeroes of $f^3$.

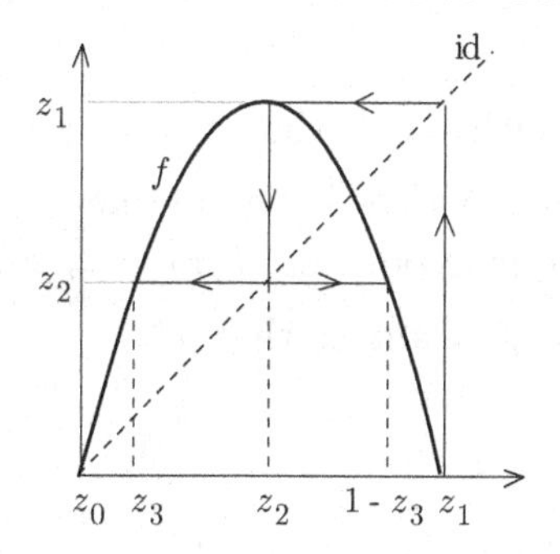


**Figure 7.4.3.** Locating zeroes of $f^3$.

## Zeroes converging to the origin

**7.4.2 Lemma** *Let $f : [0, 1] \to [0, 1]$ be a one-hump mapping and let $c \in (0, 1)$ be its critical point. If*

$$f(x) > x \quad \text{for} \quad 0 < x < c$$

*then there is a decreasing sequence of positive numbers $\{z_n\}_{n=1}^{\infty}$ converging to 0 with $z_1 = 1, f(z_{n+1}) = z_n$ and hence $f^n(z_n) = 0$.*

*Proof:* The graph of a typical $f$ of this sort on $[0, c]$ is shown below.

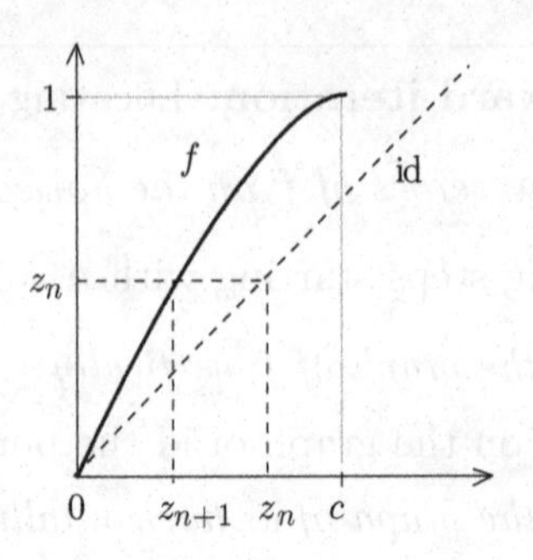


**Figure 7.4.4**

By definition of a one-hump mapping we can choose $z_1 = 1,\ z_2 = c$. Suppose we have chosen positive

$$z_1 = 1 > z_2 = c > z_3 > \cdots > z_n$$

for some $n \geq 2$ in such a way that $f(z_{i+1}) = z_i$ for every $i < n$.

By the Intermediate Value Theorem there is an $x < c$ such that $f(x) = z_n$ and by our hypothesis $f(x) = z_n > x$.

Choosing $z_{n+1} = x$ guarantees that $f(z_{n+1}) = z_n$ and $z_{n+1} < z_n$, so we now have

$$z_1 > z_2 > \cdots > z_{n+1}$$

with $f(z_{i+1}) = z_i$ for every $i < n + 1$.

Continuing in this way, we obtain a sequence $\{z_n\}$ which is monotonically decreasing and bounded below by 0. Hence there is a number $\ell \geq 0$ such that

$$\begin{aligned}
\ell &= \lim_{n\to\infty} z_n \\
&= \lim_{n\to\infty} f(z_{n+1}) && \text{by backwards iteration} \\
&= f(\lim_{n\to\infty} z_{n+1}) && \text{by continuity of } f \\
&= f(\ell).
\end{aligned}$$

Thus $\ell$ is a fixed point of $f$. But since $f >$id on $(0, c)$ there is only one fixed point for $f$, namely 0. Hence $\ell = 0$ ■

## Exercises 7.4

7.4.1. Prove the converse of Lemma 7.4.1.

7.4.2. Locate the zeroes of $f^4$ where $f$ is as shown in Figure 7.4.2.

7.4.3. Let $T : [0,1] \to [0,1]$ be the tent map.

(a) Use graphs of $T^n$ given in Section 7.2 to guess the zeroes of $T^n$ for $n = 1, 2, 3, 4$.

(b) Use the graph of $T$ to find the zeroes of $T^n$ for $n = 1, 2, 3, 4$.

(c) From your answer to part (a) and (b) guess a formula for the zeroes of $T^n$ for each $n \in \mathbb{N}$.

7.4.4. Let $T : [0,1] \to [0,1]$ be the tent map. You should have guessed that the set of zeroes of $T^n$ is

$$\{\tfrac{i}{2^{n-1}} : 0 \leq i \leq 2^{n-1}\}$$

(a) Prove this inductively using Lemma 7.4.1 and the formula

$$T(x) = \begin{cases} 2x & \text{if } 0 \leq x \leq \frac{1}{2} \\ 2 - 2x & \text{if } \frac{1}{2} < x \leq 1 \end{cases} \quad \text{for the tent map.}$$

(b) Deduce that, for each positive integer $n$, each hump of $T^n$ has base length $2^{-n+1}$.

(c) Deduce from part (b) and Lemma 7.2.5 that the tent map has wiggly iterates.

## Additional reading for Chapter 7

For material on composition of graphs see Section 3.6 of [De2] and Secions 4.2, 4.3, 4.13 of [PMJPSY].

## References

[De2] Robert L. Devaney, *Chaos, Fractals and Dynamics, Computer Experiments in Mathematics,* Addison-Wesley, New York, 1990.

[PMJPSY] Hans-Otto Peitgen, Evan Maletsky, Hartmut Jürgens, Terry Perciante, Dietmar Saupe, and Lee Yunker, *Fractals for the Classroom: Strategic Activities Volume Two,* Springer-Verlag, 1992. (two volumes)

# 8

# SENSITIVE DEPENDENCE

*When a difference equation is used to model a real world problem, the particular solution of interest is specified by an initial value, which is fed into the computer to start the iteration. The initial value, however, is usually known only approximately. Errors can arise from the limited precision of the measuring instruments and also from the limited precision to which a computer accepts numbers. There are thus two initial values to consider: the 'true" initial value $x_0$ and the approximate value $y_0$, with which the computer begins its calculation of the iterates.*

*The scientist investigating the real world problem tries to ensure that the approximate initial value $y_0$ is as close to the true initial value $x_0$ as possible. To make valid long term predictions, the scientist would need to know, furthermore, that the two solutions stay close together over many iterations. It is now known, however, that even for very simple difference equations, the two solutions can diverge so rapidly that long term predictions are impossible.*

*This rapid divergence of solutions, which are close together initially, is called* sensitive dependence on initial conditions *or* sensitive dependence *or just simply* sensitivity. *When present in a dynamical system it makes long term predictions impossible and hence is regarded as one of the key features of 'chaotic' behaviour.*

## 8.1 DIVERGING ITERATES

In this section we show how rapidly two sequences of iterates of the logistic map $Q_4$ can diverge from each other.

Each part (a), (b) and (c) of Figure 8.1.1 shows a pair of initial points $x_0$ and $y_0$ together with the graphs of their iterates. Solid line segments join the iterates of $x_0$ and dashed segments join those of $y_0$.

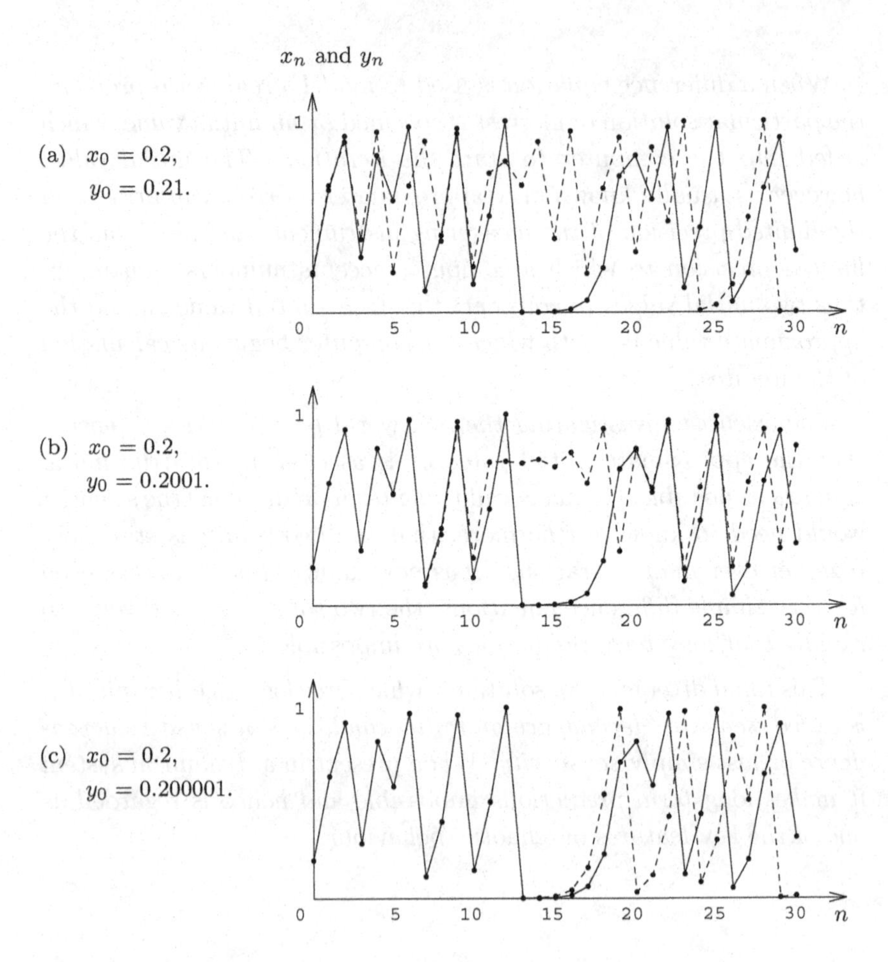


**Figure 8.1.1.** Graphs of iterates of $x_0$ and of $y_0$ . In each case (a), (b) or (c), the two sequences stay close initially, but then diverge away from each other. They diverge even when the initial points are very close, although the divergence takes longer to occur.

## Graphs of differences

It is easier to study how the iterates diverge if we graph the difference between the iterates, $|x_n - y_n| = |f^n(x_0) - f^n(y_0)|$, against $n$.

These graphs are plotted below in Figure 8.1.2 for each of the cases (a), (b) and (c). These graphs confirm that the iterates eventually diverge and that they take longer to diverge when $x_0$ is close to $y_0$.

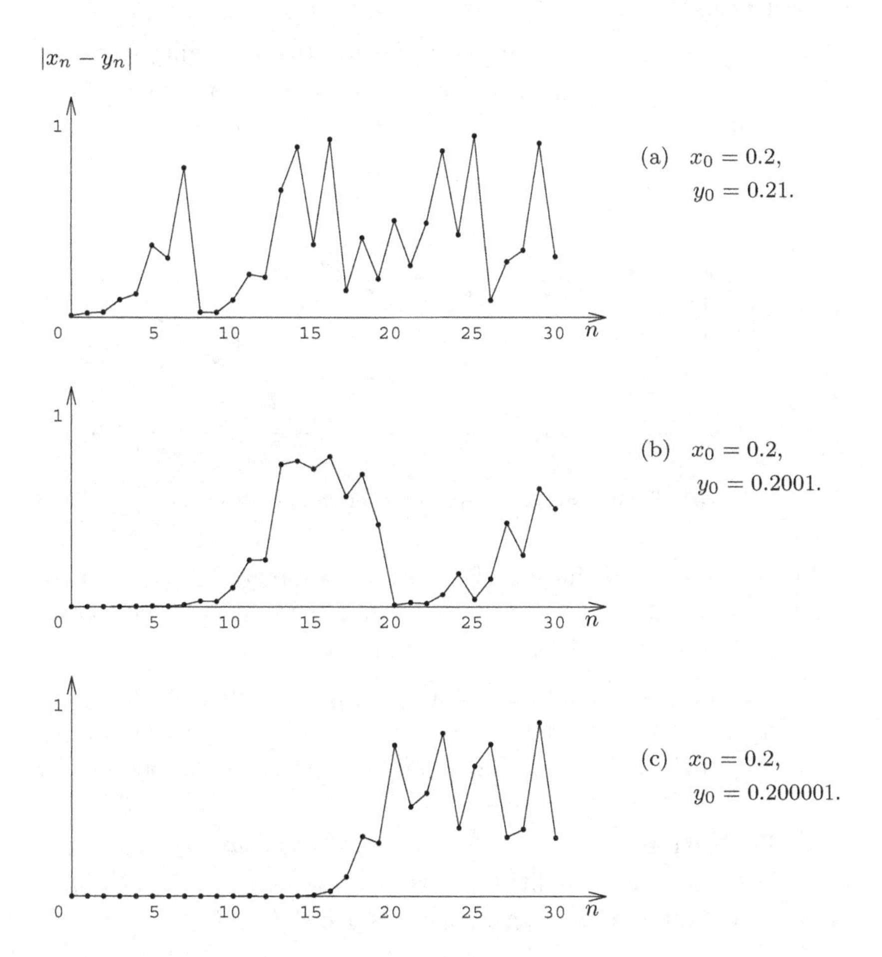


**Figure 8.1.2** Graphs of $|x_n - y_n|$ against $n$.
In case (a) the iterates separate at the first iteration.
In case (b) they separate at the seventh iteration.
In case (c) they separate at the fifteenth iteration.

## 8.2 DEFINITION OF SENSITIVITY

The previous section showed some diverging of iterates of the logistic mapping. We now show graphs of two hypothetical sequences of iterates of a mapping $f : [0,1] \to [0,1]$. The graphs will be used to provide further discussion of the idea of diverging iterates and to introduce some notation which is used in the definition of sensitive dependence.

In Figure 8.2.1, the initial points for the two diverging sequences of iterates are $x_0$ and $y_0$. These points are both in an open interval $I$. By making $I$ small we can ensure that $y_0$ is close to $x_0$.

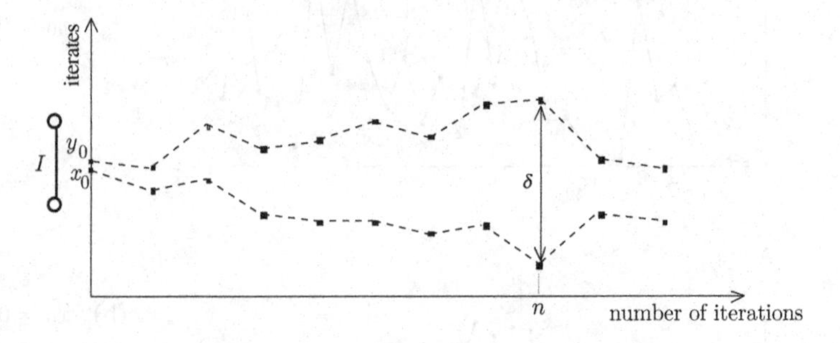


**Figure 8.2.1.** Graphs of diverging iterates.

In the following definition of sensitivity, we regard both the point $x_0$ and the number $\delta > 0$ as fixed. After $n$ iterations, the iterates become separated by at least the distance $\delta$.

As the interval $I$ becomes smaller, the point $y_0$ moves closer to $x_0$ (and we expect $n$ will get larger). Before reading the definition, you should recall the definition of an open-in-$S$ interval from Section 5.2.

**8.2.1 Definition** A map $f : [0,1] \to [0,1]$ has *sensitive dependence* at $x$ if the following condition holds for some $\delta > 0$: For each open-in-$[0,1]$ interval $I$ containing $x$, there is a $y \in I$ and an $n \in \mathbb{N}$ such that

$$|f^n(x) - f^n(y)| \geq \delta.$$

The number $\delta$ is called a *sensitivity constant* at $x$ for the mapping. ∎

From this definition it follows that if a number $\delta$ is a sensitivity constant for the mapping then so also is any smaller positive number. It is clear also that the sensitivity constant cannot exceed the length of the interval $[0, 1]$ as this interval must contain all the iterates.

**Quantifiers**

Formally, Definition 8.2.1 consists of an inequality, involving a number of variables preceded by quantifiers. To discuss the formal structure of the definition we use the standard quantifier symbols:

$\forall$ which is an abbreviation for the quantifier *for all*, and

$\exists$ which is an abbreviation for *there exists* or *there is.*

Besides using the notation '$x \in I$' to mean that $x$ is in the interval $I$, we shall also use the notation '$I \ni x$' to mean that the interval $I$ contains $x$.

When written in terms of these symbols, the definition that $f$ has sensitive dependence at $x$ reads as follows:

$$(\exists\delta > 0)(\forall I \ni x)(\exists y \in I)(\exists n \in \mathbb{N}) \quad |f^n(x) - f^n(y)| \geq \delta \qquad (1)$$

where $I$ denotes an open-in-$[0, 1]$ interval.

We shall later use the statement (1) to set out proofs that particular functions have sensitive dependence at $x$ and also to set out proofs that other functions do not have it.

The following definition is useful when we want to study mappings which show sensitive dependence at *every* point of their domains.

**8.2.2 Definition** For $f : [0, 1] \to [0, 1]$ to be *sensitive dependent everywhere* means that $f$ is sensitive dependent at each $x \in [0, 1]$ with the same sensitivity constant $\delta$. ∎

In the above definitions, the interval [0,1] can be replaced by any set $S \subseteq \mathbb{R}$. The definition is of most interest, however, when $S$ is a closed bounded[1] subset of $\mathbb{R}$.

## Exercises 8.2

All exercises in this section refer to a mapping

$$f : [0,1] \to [0,1].$$

8.2.1. Let $x \in [0,1]$. Of the numbers $0.5$, $0$, $2$, $-0.5$, which ones cannot be sensitivity constants at $x$ for $f$ ?

8.2.2. Which of the following statements are correct?

(a) If $\delta$ is a sensitivity constant at $x$ for $f$, then so is $2\delta$.

(b) If $\delta$ is a sensitivity constant at $x$ for $f$, then so is $\delta/2$.

(c) If $\delta$ is not a sensitivity constant at $x$ for $f$, then neither is $2\delta$.

8.2.3. An environmental scientist wants to predict the size of a population by using a difference equation

$$x_{n+1} = f(x_n)$$

with the population scaled so as to lie in $[0,1]$. The mapping $f$ is known to be sensitive dependent at the initial value $x_0$ which the scientist wishes to use, with sensitivity constant $\delta > 0$.

In which of the following cases should the scientist definitely start worrying about the accuracy of the predictions?

(i) $\delta = 10^{-15}$, (ii) $\delta = 0.5$, (iii) $\delta = 1$.

[1] If the set $S$ is unbounded, say $S = \mathbb{R}$, then both the errors and the iterates can increase beyond all bounds. Hence it is not the *absolute error* $|f^n(x) - f^n(y)|$ which is of interest, but the *relative error*, which is the absolute error divided by the size of the iterate.

8.2.4. In the Definition 8.2.1 of sensitive dependence, is the inequality

$$|f^n(x_0) - f^n(y_0)| \geq \delta$$

required to hold for

(a) *all* values of $y_0$ close to $x_0$ ?

(b) only *some* values of $y_0$ arbitrarily close to $x_0$ ?

8.2.5. Keeping an eye on Figure 8.2.1 (as a pictorial prompt) write out in full the definition of the statement that $f$ *has sensitive dependence at* $x_0$.

The following statements are referred to in Exercises 6 and 7.

(I) There is a $\delta > 0$ such that for each $I \ni x$ there is a $y \in I$ and an $n \in \mathbb{N}$ such that

$$|f^n(x) - f^n(y)| \geq \delta.$$

(II) There is a $\delta > 0$ and a sequence $\{y_k\}_{k=1}^{\infty}$ converging to $x$ and for each $y_k$ there is an $n \in \mathbb{N}$ such that

$$|f^n(x) - f^n(y_k)| \geq \delta.$$

(III) There is a $\delta > 0$ and an $n \in \mathbb{N}$ such that for all $I \ni x$ there is a $y \in I$ such that

$$|f^n(x) - f^n(y)| \geq \delta.$$

As usual, $I$ is an open-in-$[0, 1]$ interval.

8.2.6. Write out each of the statements (I), (II) and (III) given in the box above, using the quantifier symbols.

8.2.7. Consider the three statements (I), (II), and (III) given in the box before the previous exercise.

(a) Which of these statements is the same as the definition that $f$ has *sensitive dependence* at $x$?

(b) Which of these statements means that there is an $n \in \mathbb{N}$ such that $f^n$ is *discontinuous* at $x$?

(c) What does the remaining statement mean?

[Hint for (c). See Exercise 9 below.]

8.2.8. The mapping $f$ is said to be *expansive* at $x \in [0, 1]$ if the following condition holds for some $\delta > 0$: for every $y \in [0, 1]$ with $y \neq x$ there is an $n \in \mathbb{N}$ such that

$$|f^n(x) - f^n(y)| \geq \delta.$$

(a) Write out the definition that $f$ is expansive at $x$ in terms of quantifier symbols.

(b) Explain in your own words how *expansiveness* differs from *sensitivity*.

8.2.9. Let $Y$ be any subset of $\mathbb{R}$ and let $x \in \mathbb{R}$.

Assume: *For every open-in-$Y$ interval $I$ containing $x$ the set $I$ also contains a point of $Y$.*

Deduce: *there is a sequence $(y_k)_{k=1}^{\infty}$ in $Y$ converging to $x$.*

Prove also the converse of this result.

8.2.10. Suppose we weakened the requirement that the interval $I$ in the definition of sensitive dependence be open-in-$[0, 1]$, instead requiring only that $I$ be an interval contained in $[0, 1]$.

Show that under this revised definition, the map

$$f : [0, 1] \to [0, 1] : x \mapsto \begin{cases} 0 & \text{if } x \in [0, \frac{1}{2}] \\ Q_4^2(x) & \text{if } x \in (\frac{1}{2}, 1] \end{cases}$$

does *not* have sensitive dependence at $\frac{1}{2}$.

[Hint: Consider intervals having $\frac{1}{2}$ as right endpoint.]

## 8.3 USING THE DEFINITION

Recall that, when written with symbols for the quantifiers, the definition that $f$ has sensitive dependence at $x$ is

$$(\exists\delta > 0)(\forall I \ni x)(\exists y \in I)(\exists n \in \mathbb{N}) \quad |f^n(x) - f^n(y)| \geq \delta.$$

where $I$ denotes an open-in-$[0,1]$ interval.

The pattern of the quantifier symbols suggest the overall structure of any proof of sensitivity. Thus, for example, the first quantifier $\exists\delta > 0$ tells us that we must *give an example* of a number $\delta$ which satisfies the conditions specified. Hence:

To prove from the definition that $f$ is sensitive at $x$

*you must choose a sensitivity constant $\delta$*

such that

*for each $I$ containing $x$ there is a $y \in I$ and*

an integer $n \geq 0$ such that

$$|f^n(x) - f^n(y)| \geq \delta. \tag{2}$$

The proof thus breaks up initially into two steps:

STEP 1: choosing a number $\delta > 0$ for which (2) holds;

STEP 2: verifying that the chosen $\delta$ works.

Choosing a sensitivity constant $\delta$ is usually the hardest part of the proof since it involves looking ahead to see what conditions $\delta$ has to satisfy. The working involved in the choice of $\delta$ has the status of an *exploration,* enclosed within exploration signs, and followed by an indication of our choice of $\delta$:

*Choose* $\delta = \ldots$ .

The verification that the chosen $\delta$ works involves checking (2), which begins with the quantifier $\forall I$, indicating that what follows holds *for every* $I \ni x$. Hence we begin the verification by saying:

*Let $I$ be an open-in-*$[0,1]$ *interval containing $x$.*

These ideas are illustrated in the following example.

**8.3.1 Example** *Show that the mapping $f : [0,1] \to [0,1]$ with $f(x) = x^2$ has sensitive dependence at 1.*

*Solution:*

To find a $\delta > 0$ such that for each open-in-$[0,1]$ interval $I$ containing 1, there is a $y \in I$ and an integer $n$ such that

$$|f^n(y) - f^n(1)| \geq \delta.$$

Note that iteration gives $f^n(x) = x^{2^n}$ for $x \in [0,1]$. Hence under $f$ the sequence of iterates of 1 is

$$(1, 1, 1, 1, \dots)$$

and that of $y$ is

$$(y, y^2, y^4, y^8, \dots) .$$

These two sequences are plotted below on the same graph, for a typical $y \in (0,1)$ .

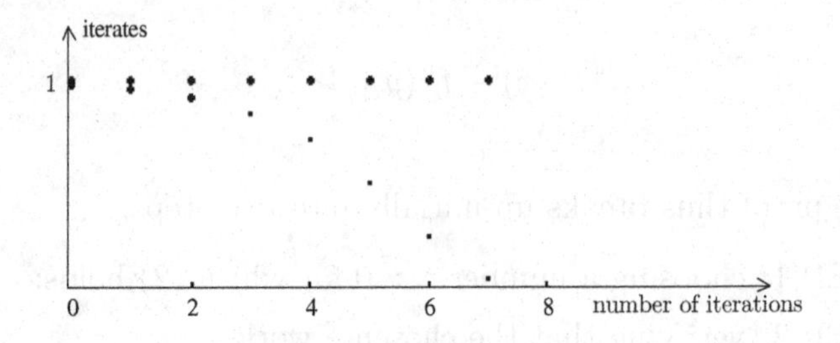


**Figure 8.3.1.** Diverging iterates of $x_0 = 1$ and $y_0 = .99$.

But since $0 \leq y < 1$ it follows that $y^{2^n} \to 0$ as $n \to \infty$. This puts $f^n(y)$ down close to the horizontal axis when $n$ is big. t Hence we can pick $\delta$ to be any number between 0 and 1.

Choose $\delta = \frac{1}{2}$. Let $I$ be an open-in-$[0,1]$ interval containing 1.
Choose $y$ to be any number in $I$ such that $y \neq 1$.
Then choose $n$ so large that $y^{2^n} < \frac{1}{2}$. This gives $|1 - y^{2^n}| \geq \frac{1}{2}$.
Hence $|f^n(y) - f^n(1)| \geq \delta$. Thus $f$ has sensitive dependence at 1. ∎

## Negation

In order to understand a concept it is often helpful to look at its opposite, obtained by negating the definition of the original concept. The opposite concept to sensitivity is obtained by negating the definition that a mapping $f$ has sensitive dependence at a point $x$; that is, by negating the statement

$$(\exists\delta > 0)(\forall I \ni x)(\exists y \in I)(\exists n \in \mathbb{N}) \quad |f^n(x) - f^n(y)| \geq \delta$$

where $I$ denotes an open-in-$[0,1]$ interval. The formal procedure for negating a quantified statement like this is to first turn all the $\exists$ symbols into $\forall$ symbols and vice-versa; then replace the inequality $|f^n(y) - f^n(x)| \geq \delta$ by its negation. This gives:

$f$ is *not* sensitive dependent at $x$ if

$$(\forall\delta > 0)(\exists I \ni x)(\forall y \in I)(\forall n \in \mathbb{N}) \quad |f^n(x) - f^n(y)| < \delta$$

where $I$ denotes an open-in-$[0,1]$ interval. Intuitively, this means: *given any $\delta > 0$ the iterate of every point $y$ will stay within a distance $\delta$ of the iterate of $x$ provided that $y$ is close enough to $x$.*

To prove from the definition that $f$ is not sensitive at $x$:

STEP 1: Let $\delta > 0$

*You must now choose an open-in-$[0,1]$ interval $I$ containing $x$ such that for every $y \in I$ and every $n$, $|f^n(y)-f^n(x)| < \delta$.*

Thus the proof then involves:

STEP 2: An exploration to find a suitable interval $I$ and

STEP 3: A verification that the chosen $I$ works.

The following example shows how to prove the negation.

**8.3.2 Example** *Prove that $f : [0,1] \to [0,1]$ with $f(x) = \frac{1}{2}x$ does not have sensitive dependence at 0.*

*Solution:*

Let $\delta > 0$. To show that for some open-in-$[0,1]$ interval $I$ containing 0, for every $y \in I$ and every $n$

$$|f^n(y) - f^n(0)| < \delta.$$

To find $I$ such that for every $y \in I$ and every $n$

$$f^n(y) < \delta.$$

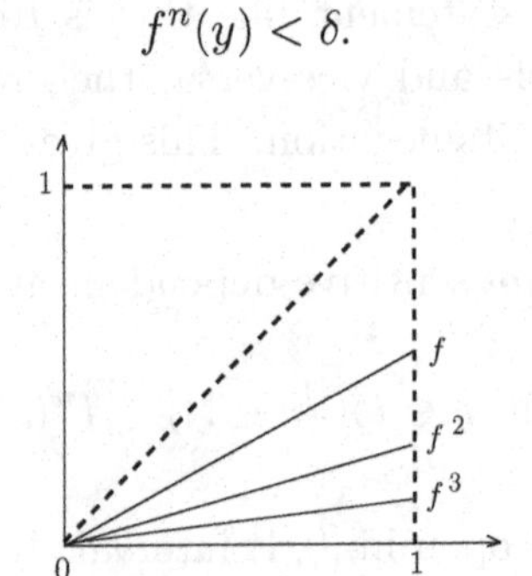


**Figure 8.3.2.** Graphs of $f$, $f^2$ and $f^3$.

The graphs of all the iterates of $f$ lie below the graph of $f$.
Hence we need only to satisfy the inequality in the case $n = 1$.
It will then be satisfied for all $n > 1$.
Thus the inequality to be satisfied is $f(y) < \delta$; that is $\frac{1}{2}y < \delta$.
This must hold for all $y \in I$.
Hence choose $I$ to have length at most $2\delta$.

Let $I$ be an interval of length[2] min($2\delta$,1) and let $n \in \mathbb{N}$. Then for $y \in I$,

$$|f^n(y) - f^n(0)| = \left(\frac{1}{2}\right)^n y < \frac{1}{2}y < \delta.$$

Thus $f$ does not have sensitive dependence at 0. ■

[2] The notation $\min(a, b)$ denotes the smaller of the two numbers $a$ and $b$. For example: $\min(2,4) = 2$, $\min(5,5) = 5$ and $\min(2,-4) = -4$.

## Graphical interpretation

As the method of solution for Example 8.3.2 in the text suggests, sensitive dependence can be interpreted in terms of the graphs of the higher iterates of $f$.

For example, let $f : [0,1] \to [0,1]$ with

$$f(x) = x^2.$$

As we showed in Example 8.3.1, this map has sensitive dependence at the point 1 of its domain. We might expect this to happen, given that the graph of a typical $n$th iterate appears as in Figure 8.3.3.

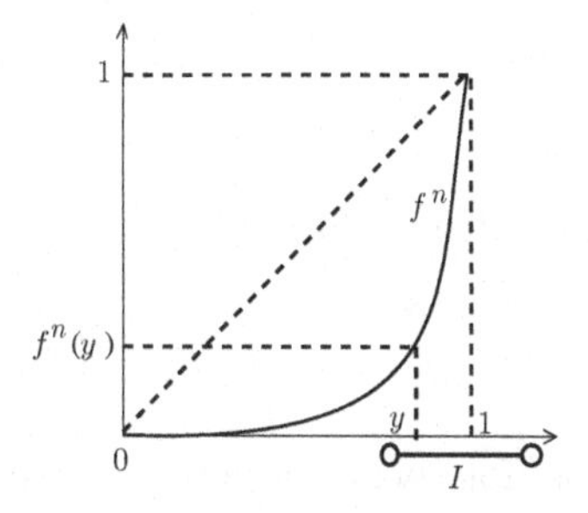


**Figure 8.3.3**

The intuitive idea is that *the graph rises very steeply* over the interval between $y$ and 1 when $y$ is close to 1. This produces *a large separation* between $f^n(y)$ and $f^n(1)$. This leads immediately to sensitive dependence.

The idea that the steepness of a graph controls its increase in height is expressed precisely by the Mean Value Theorem from calculus, which we state below.

**Mean Value Theorem.** *If $g$ is differentiable on $[a,b]$, then there is an $x \in (a,b)$ such that*

$$g(b) - g(a) = g'(x)(b-a).$$

## Exercises 8.3

For all the exercises in this section we let $f : [0,1] \to [0,1]$.

8.3.1. Show that in Example 8.3.1 you could have chosen the sensitivity constant $\delta$ to be $\frac{3}{4}$.

8.3.2. Let $f(x) = x^3$. Prove that $f$ has sensitive dependence at 1.

8.3.3. Let $g : [-1,1] \to [-1,1]$ with $g(x) = x^3$. Guess two points at which $g$ is sensitive dependent and then prove your guess.

8.3.4. Use the previous exercise to find a mapping $f : [0,1] \to [0,1]$ which is sensitive dependent at two points.

8.3.5. Let $f(x) = x^2$. Prove that $f$ does not have sensitive dependence at 0.

8.3.6. Let $f(x) = x$.

(a) Prove that $f$ is not sensitive dependent at 1.

(b) Prove that $f$ is not sensitive dependent anywhere.

8.3.7. Let $f(x) = \frac{1}{3}x$. Prove that $f$ is not sensitive dependent anywhere.

8.3.8. Let $|f'(x)| < 1$ for all $x \in [0,1]$. Prove that $f$ is not sensitive dependent anywhere.

[Hint: Show that $|(f^n)'(x)| < 1$ for all $x \in [0,1]$ and then use the Mean Value Theorem.]

8.3.9. Show that if $f : [0,1] \to [0,1]$ is differentiable, it is impossible to have $f'(x) > 1$ for all $x \in [0,1]$.

[Hint: Use the Mean Value Theorem.]

## 8.4 SENSITIVITY AND CHAOS

> *"Much of chaos as a science is connected with the notion of 'sensitive dependence on initial conditions'. Technically, scientists term as 'chaotic' those non-random complicated motions that exhibit a very rapid growth of errors that, despite perfect determinism, inhibits any pragmatic ability to render accurate long-term prediction.*
>
> *The property of sensitivity is central to chaos. Sensitivity, however, does not automatically lead to chaos."*
>
> – *Peitgen*

Although the association between chaotic behaviour and sensitive dependence is often stressed in the literature, the mapping

$$f : [0,1] \to [0,1] \text{ with } f(x) = x^2 \tag{1}$$

manages to have sensitive dependence at 1 without appearing to be very chaotic. Its behaviour in fact seems quite regular and predictable. The reason for this seems to be that it is sensitive dependent at *only one point* of its domain.

Unpredictable and hence chaotic behaviour is expected, however, when sensitivity occurs at *every point* of the domain of a mapping.

> *"It would seem that for this situation to occur (small cause producing a big effect), one needs an exceptional state at time zero. The opposite is true: many physical systems exhibit sensitive dependence on initial condition for arbitrary initial condition. This is somewhat counterintuitive, and it has taken some time for mathematicians and physicists to understand well how things happen."*
>
> – *Ruelle*

To achieve sensitivity at every point in its domain it would seem that the graphs of the higher iterates must rise very steeply above every small interval in the domain. In the next chapter we will give examples of mappings for which this happens.

## Additional reading for Chapter 8

The concept of sensitive dependence was originally introduced in [Guc], although this paper is somewhat difficult to read. Gentler introductions to the mathematics are given in Section 1.8 of [Del], Chapter 9 of [Ho], Section 2.1 of [Gul] and Section 10.1 of [PJS]. Discussions of the consequences of sensitive dependence for predictive sciences are given in [Gl], [PJS] and[Ste].

## References

[Del] Robert L. Devaney, *An Introduction to Chaotic Dynamical Systems: Second Edition,* Addison-Wesley, Menlo Park California, 1989.

[Gl] James Gleick, *Chaos: Making a New Science,* Sphere Books, London, 1988.

[Guc] John Guckenheimer, "Sensitive Dependence on Initial Conditions for One-Dimensional Maps", *Communications in Mathematical Physics,* **70** (1973), 133-160.

[Gul] Denny Gulick, *Encounters with Chaos,* McGraw-Hill, Inc., 1992.

[Ho] Richard Holmgren, *A First Course in Discrete Dynamical Systems* Springer-Verlag, 1994.

[PJS] Hans-Otto Peitgen, Hartmut Jürgens and Dietmar Saupe, *Chaos and Fractals, New Frontiers of Science*, Springer-Verlag, New York, 1992.

[Ste] Ian Stewart, *Does God Play Dice?* Penguin Books, Middlesex, 1989.

[Ru2] David Ruelle, *Chance and Chaos,* Penguin 1991.

# 9

# INGREDIENTS OF CHAOS

*In the previous chapter we introduced the idea of a mapping with sensitive dependence and mentioned the relevance of this idea to the study of chaotic behaviour.*

*For simplicity, emphasis was given to mappings showing sensitive dependence at a single point of their domains. In this chapter, however, the emphasis shifts to mappings which show sensitive dependence at every point in their domains.*

*Examples given in the last chapter suggest that sensitive dependence of a mapping occurs at a point $x$ of its domain with the following property: graphs of higher iterates rise steeply over each small interval containing the point $x$. Hence for a mapping to be sensitive everywhere the graphs of higher iterates must rise steeply above each small interval in the domain.*

*In this chapter we prove that mappings with wiggly iterates have sensitive dependence everywhere. The higher iterates of these mappings wiggle up and down between 0 and 1. Above every small interval in the domain, we can find an iterate with slope as large as we please.*

*The wiggly behaviour of the iterates will be shown to imply that the mapping has not only*

*(a) sensitive dependence everywhere, but also*

*(b) transitivity, and*

*(c) a dense set of periodic points.*

*In a popular definition, due to the U.S. mathematician Robert Devaney, these three properties are taken as the essential ingredients of chaos.*

## 9.1 SENSITIVITY EVERYWHERE

The basis for nearly everything we do in this chapter is the Spike Lemma (Lemma 7.2.4). It shows that graphs of higher iterates of mappings with wiggly iterates rise steeply above each small interval in the domain. We restate the lemma here for convenience.

**Spike Lemma** *Let $f : [0,1] \to [0,1]$ have wiggly iterates. For every interval $I \subseteq [0,1]$ there is a hump, of some $f^n$, whose base is contained in $I$.* ■

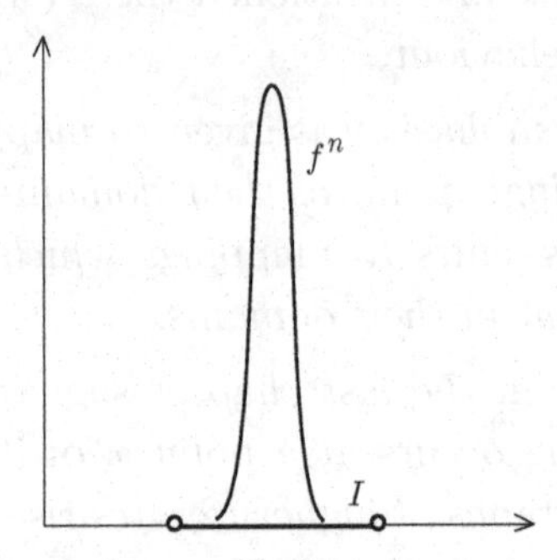


**Figure 9.1.1.** A hump given by the Spike Lemma.

Since we can get an iterate with slope as large as we please over any given interval, we guess that mappings with wiggly iterates have sensitivity everywhere on the interval $[0,1]$. This suggests the following theorem.

### 9.1.1 Lemma (Wiggly implies sensitive)

*If $f : [0,1] \to [0,1]$ has wiggly iterates, then $f$ has sensitive dependence everywhere with sensitivity constant $\frac{1}{2}$.*

*Proof:* Assume that the mapping $f$ has wiggly iterates.

We shall deduce that $f$ has sensitive dependence everywhere. To do this, we verify that the definition of sensitive dependence (Definition 8.2.2) is satisfied; that is, we verify that

$$(\exists \delta > 0)(\forall x \in [0,1])(\forall I \ni x)(\exists y \in I)(\exists n \in \mathbb{N}) \quad |f^n(x) - f^n(y)| \geq \delta$$

where $I$ denotes an open-in-$[0,1]$ interval.

We adopt the suggested choice $\delta = \frac{1}{2}$ for the sensitivity constant.

Let $x$ be any point in $[0, 1]$.

Let $I$ be any open-in-$[0, 1]$ interval containing $x$.

We have to find $y \in I$ and $n \in \mathbb{N}$ such that $|f^n(x) - f^n(y)| \geq \frac{1}{2}$.

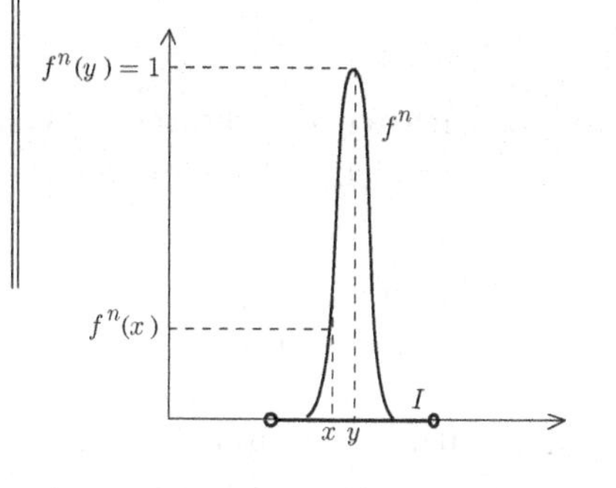


**Figure 9.1.2(a)**

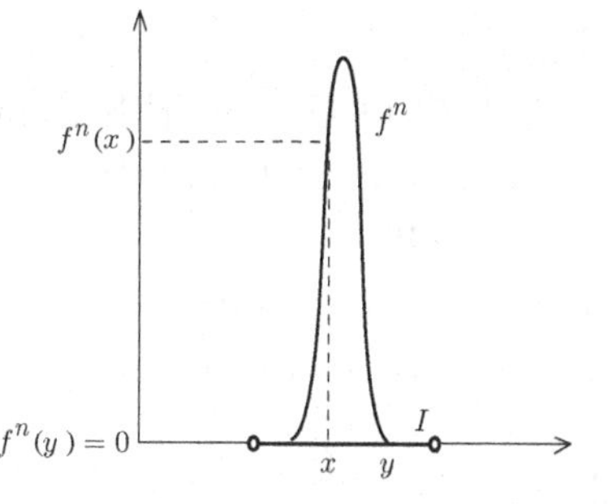


**Figure 9.1.2(b)**

By the Spike Lemma there is an $n$ such that the base of some hump of $f^n$ is contained in $I$.

Let $n$ be chosen as in the exploration above.

If $f^n(x) \leq \frac{1}{2}$, choose $y \in I$ so that $f^n(y) = 1$. See Figure 9.2.1(a).

If $f^n(x) > \frac{1}{2}$, choose $y \in I$ so that $f^n(y) = 0$. See Figure 9.2.1(b).

Hence, in both cases there is a $y \in I$ such that

$$|f^n(y) - f^n(x)| \geq \frac{1}{2}.$$

Thus maps with wiggly iterates have sensitive dependence on initial conditions everywhere, with sensitivity constant $\frac{1}{2}$. ∎

## Exercises 9.1

9.1.1. Show from the *Wiggly Implies Sensitive Lemma*, together with results from earlier chapters, that both the logistic mapping $Q_4$ and the tent mapping $T_4$ have sensitive dependence everywhere.

9.1.2.

(a) Let $f : [0,1] \to [0,1]$ be a mapping with wiggly iterates. Give an example of a point $x \in [0,1]$ for which $f$ has sensitivity constant $\delta = 1$.

(b) Prove the validity of your answer to part (a). Set your proof out carefully, following the one given above in the text.

9.1.3. Let $f : [0,1] \to [0,1]$ be a mapping with wiggly iterates. Show that there is a dense set of points in the interval $[0,1]$ at which $f$ has sensitivity constant $\delta = 1$.

[Hint. See the discussion of the zeroes of iterates of mappings with wiggly iterates in Section 7.3.]

9.1.4. A mapping $f : [0,1] \to [0,1]$ is said to be *spreading* if for each interval $I \subseteq [0,1]$ there is an $n \in \mathbb{N}$ such that $f^n(I) \supseteq [0,1]$.

(a) Which result given in the text shows that a mapping with wiggly iterates is spreading?

(b) Prove that if $f$ is one-hump mapping and spreading then it has wiggly iterates.

## 9.2 DENSE ORBITS AND TRANSITIVITY

Although sensitivity is generally agreed to be one of the most important ingredients in chaotic behaviour, there are other ingredients which are also very relevant. Sensitivity refers to the behaviour of *two orbits* which diverge away from each other as iteration proceeds.

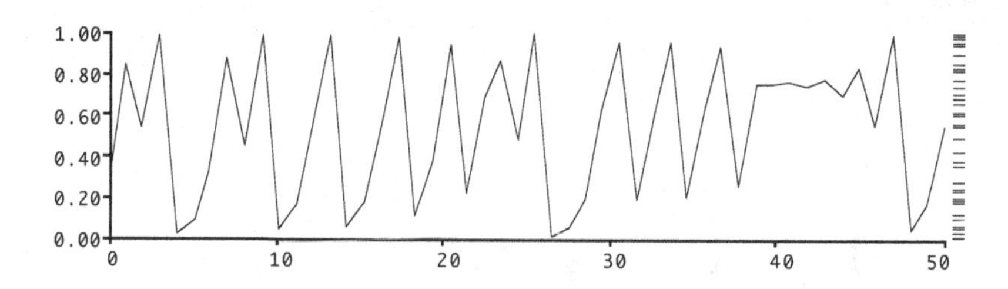


**Figure 9.2.1.** The orbit of the point $x_0 = 0.3$ under the logistic mapping $Q_4$.

On the other hand, the presence of chaos is often inferred from the behaviour of *a single orbit* which wanders all over the codomain of the mapping (instead of settling down to regular, predictable behaviour). The graph of such an orbit is shown in Figure 9.2.1.

The orbit seems to be 'chaotic'. The points of the orbit have been projected onto the vertical line at the right. They appear to be filling up the interval $[0, 1]$ densely.

Thus Figure 9.2.1 suggests that we can replace the intuitive idea of a 'chaotic orbit' by the precise mathematical concept of a *dense orbit.*

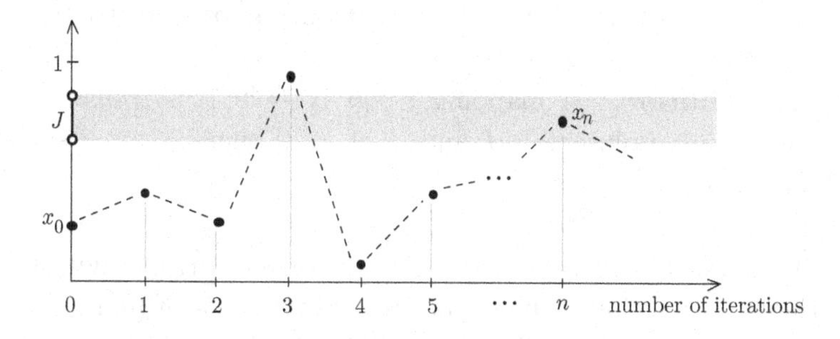


**Figure 9.2.2.** Proving an orbit is dense.

To prove that the orbit of $x_0$ is dense (see Figure 9.2.2) we show that:

*for every interval $J \subseteq [0,1]$ there is an element $x_n$ of the orbit in $J$.*

Proving the existence of a dense orbit, however, requires more Topology than we assume in this book. We therefore turn to a closely related concept known as *transitivity.* In place of the orbit of a point $x_0$, the idea is to consider the iterates of a small interval $I_0$ (as in Figure 9.2.3).

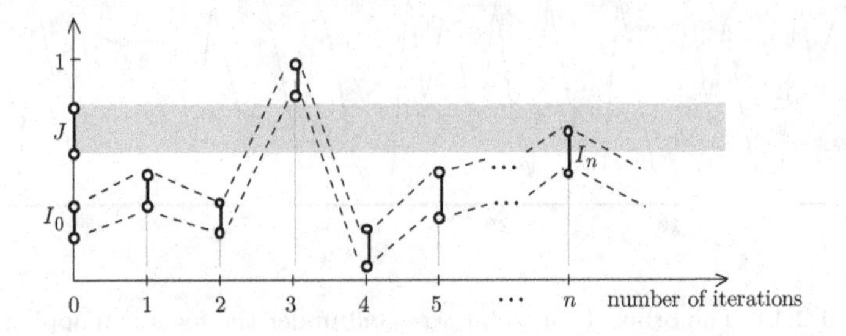


**Figure 9.2.3.** Proving a mapping is transitive.

This makes good computer sense because the numbers with a given truncation form a small interval. The definition of 'an orbit is dense', using iterates of a small interval in place of a number (see Figure 9.2.3) is then:

*for every interval $J$, some iterate of $I_0$ must intersect $J$.*

To formalize the idea of the interval $I_0$ being *small*, we require this property to hold for *every* interval $I_0$. We then say that the mapping $f$ is *transitive.*

These ideas lead to the following definition of transitivity.

**9.2.1 Definition** A mapping $f : [0,1] \to [0,1]$ is *transitive* if for every pair of subintervals $I$ and $J$ of $[0,1]$ there is an $n$ such that $f^n(I) \cap J \neq \varnothing$. ■

A further illustration of the ideas involved in transitivity is given in Figure 9.2.4. Let $I$ and $J$ be two subintervals of $[0,1]$. Let $f$ be the logistic map. To see if some iterate of $I$ intersects $J$ we plot the graph of the orbit of the endpoints of $I$ under $f$.

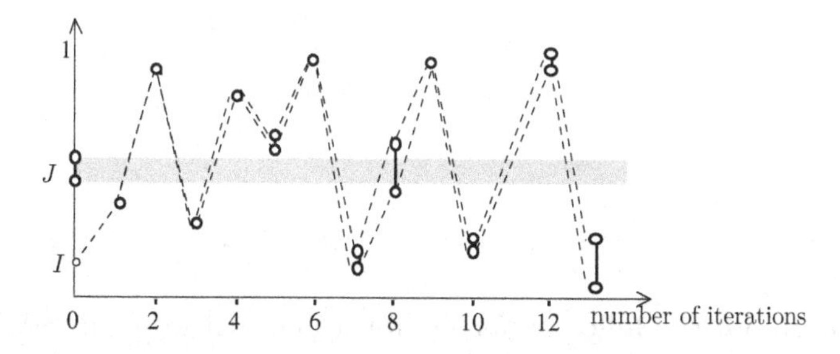


**Figure 9.2.4.** Iterates of the interval $I = [.1, .1005]$.

Given the intervals $I$ and $J$ shown in the figure, we see that the smallest positive integer for which $f^n(I) \cap J \neq \varnothing$ is $n = 8$.

Written in terms of quantifier symbols, the definition of transitivity reads:

$$(\forall I)(\forall J)(\exists n \in \mathbb{N}) \quad f^n(I) \cap J \neq \varnothing.$$

Hence the pattern of the proof that $f$ is transitive is:

Let $I$ and $J$ be some intervals.

Find an $n$ such that $f^n(I) \cap J \neq \varnothing$.

### 9.2.2 Lemma (Wiggly implies transitive)

*If $f : [0,1] \to [0,1]$ has wiggly iterates then $f$ is transitive.*

*Proof:* Let $f$ have wiggly iterates. Let $I$ and $J$ be any subintervals of $[0,1]$.

We have to find an $n$ such that $f^n(I) \cap J \neq 0$.

By the Spike Lemma there is an $n$ such that the base of some hump of $f^n$ is contained in $I$.

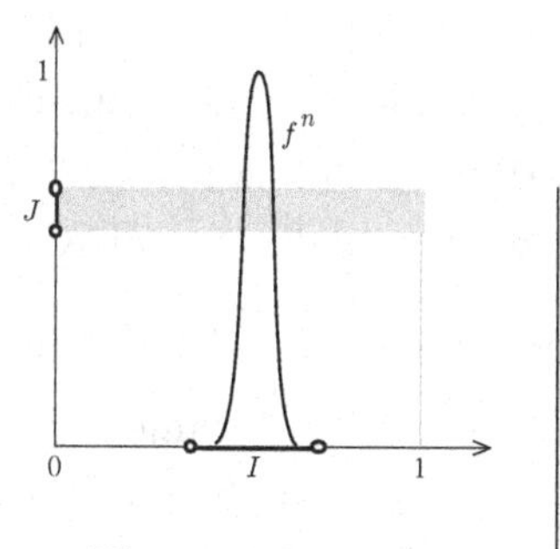


**Figure 9.2.5**

Choose $n$ such that $f^n(I) = [0, 1]$. See Figure 9.2.5.
Hence $f^n(I) \cap J \neq \varnothing$. Thus $f$ is transitive. ■

## Exercises 9.2

9.2.1. Of the following orbits for a map with domain $[0, 1]$, which (if any) are dense? Give reasons.

(a) A periodic orbit.

(b) An eventually periodic orbit.

(c) An orbit converging to a fixed point..

9.2.2. Write out from the definition the meaning of the statement that a mapping $f$ is not transitive. Show all the quantifiers. Hence describe the steps in a proof of such a statement.

9.2.3. Show that each of the following mappings is not transitive:

(a) id: $[0, 1] \to [0, 1]$;

(b) $\mathrm{id}^2 : [0, 1] \to [0, 1]$.

9.2.4. Decide whether the mapping $f : [0, 1] \to [0, 1]$ with

$$f(x) = \begin{cases} 2x & \text{if } 0 \leq x \leq 1/2 \\ 3/2 - x & \text{if } 1/2 \leq x \leq 1 \end{cases}$$

is transitive.

9.2.5. Show that if a mapping has an attracting fixed point, then the mapping is not transitive.

[Hint: It follows from Theorem 5.2.1 that $f(K) \subseteq K$ for every small enough interval $K$ containing the fixed point.]

9.2.6. Let $D$ be a dense subset of $[0, 1]$.

(a) Let $S$ be obtained from $D$ by omitting a single point, say $x_0$. Prove that $S$ is also dense in $[0, 1]$.

(b) Deduce that if $S$ is obtained from $D$ obtained by omitting a finite number of points , then $S$ is also dense in $[0, 1]$.

9.2.7. Deduce from Exercise 6 that every point of a dense orbit is an initial value for a dense orbit.

9.2.8. Prove the following theorem (using the result of Exercise 7):

*if a mapping $f : [0, 1] \to [0, 1]$ has a dense orbit, then the mapping is transitive.*

## 9.3 DENSE PERIODIC POINTS

Periodic orbits seem to lie at the opposite pole from chaotic behaviour. It may come as a surprise, therefore, to find the existence of lots of periodic points tied to the other ingredients of chaos.

**9.3.1 Lemma (Wiggly implies dense periodic points)**
*If $f : [0,1] \to [0,1]$ has wiggly iterates, then the set of all periodic points of $f$ is dense in the interval* $[0,1]$.

*Proof:* Let $f$ have wiggly iterates.
We will show that the set of periodic points of $f$ is dense in $[0,1]$.

> We want to show that every interval $I \subseteq [0,1]$ contains a periodic point of $f$.

Let $I$ be any subinterval of [0,1]. By the Spike Lemma, there is a hump of some $f^n$ over the interval $I$ as indicated in Figure 9.3.1.

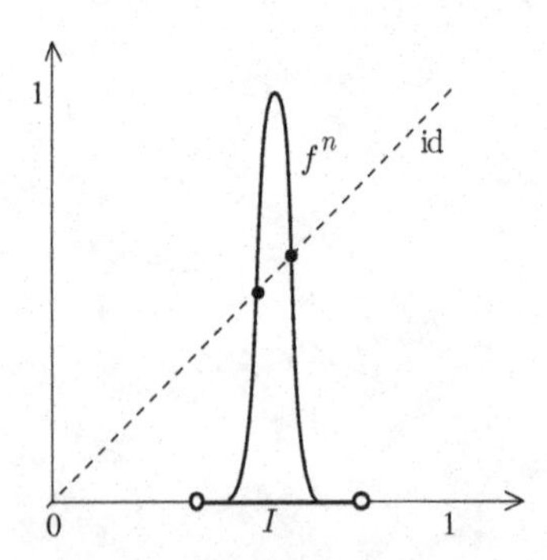


**Figure 9.3.1** The interval $I$ contains two periodic points of $f$.

This hump must intersect the graph of id in at least two points, by the Intermediate Value Theorem (see Exercise 9.3.2 ).

These points of intersection, when projected onto the domain, give two periodic points of $f^n$. Thus $f$ has two periodic points (and hence at least one such point) in $I$.

The set of periodic points is therefore dense in [0,1]. ∎

## Exercises 9.3

9.3.1. In each case sketch the graph of a symmetric one-hump mapping $f : [0,1] \to [0,1]$ which has the stated number of fixed points:

(a) two fixed points;

(b) three fixed points;

(c) four fixed points.

What is the least possible number of fixed points for a one-hump map?

9.3.2. Give three examples of mappings whose periodic points are dense in their domains. Give reasons for your answers.

9.3.3. Let $f : [0,1] \to [0,1]$ be a one-hump mapping. Suppose that $f^n$ has a hump with base contained in an interval $I$, as in Figure 9.3.1.

Thus there are points $z_i < y_i < z_{i+1} \in I$, such that

$$f^n(z_i) = 0, \qquad f^n(z_{i+1}) = 0 \qquad \text{and} \qquad f^n(y_i) = 1.$$

(a) Use the Intermediate Value Theorem to prove that $f^n - \mathrm{id}$ has at least two zeroes between $z_i$ and $z_{i+1}$.

(b) Deduce that $f^n$ has at least two fixed points in the interval $I$.

9.3.4. Suppose $p$ is an attracting fixed point for a map $f : [0,1] \to [0,1]$.

(a) Deduce from Theorem 5.2.1 that there is an open-in-[0,1] interval $I$ such that $|f^n(x) - p| < |x - p|$ for all $x \in I \setminus \{p\}$ and all $n \in \mathbb{N}$.

(b) Deduce that the fixed points of $f$ are not dense in the domain of $f$.

(c) Give an example of such a mapping.

## 9.4 WHAT IS CHAOS?

In the previous sections it was shown that mappings with wiggly iterates have:

(a) *sensitive dependence;*

(b) *transitivity* (and *a dense orbit);*

(c) *a dense set of periodic points.*

Properties (a) and (b) correspond to irregular unpredictable behaviour and hence agree with intuitive ideas about chaos.

Property (c), however, seems at odds with the previous two properties since we usually think of periodic behaviour as being regular and predictable. On the other hand, these periodic points are all unstable and this means that as soon as an orbit comes close to a periodic point, it will be pushed away somewhere else. So in practical terms, a dense set of periodic orbits does not imply orderly behaviour.

In 1989 Devaney gave a definition of chaos based on these three properties. Whatever one might think about the inclusion of property (c) in a definition of chaos, there is no doubt that Devaney's definition has caught on. It probably the most popular definition of chaos in mathematics textbooks and research papers nowadays.

**9.4.1 Definition** Let $V$ be an interval. We say that $f : V \to V$ is *chaotic* on $V$ if:

(a) $f$ has sensitive dependence on initial conditions;

(b) $f$ is transitive;

(c) periodic points are dense in $V$. ∎

Lemmas in previous sections show that having wiggly iterates implies each of the conditions (a), (b) and (c). Hence, combining these lemmas and using the definition of chaotic behaviour proves the following theorem.

**9.4.2 Theorem (Wiggly implies chaotic)**

*If $f : [0, 1] \to [0, 1]$ has wiggly iterates then $f$ is chaotic on $[0, 1]$.* ∎

As we showed, in Chapter 7, the tent map and the logistic map with $\mu = 4$ have wiggly iterates. Hence Theorem 9.4.2 has the following corollary:

*The logistic mapping $Q_4$ and the tent mapping $T_4$ are chaotic on the interval* $[0, 1]$.

As we have already mentioned, Devaney's definition is applicable in a wider context than just mappings of an interval. All that is required is that $V$ be any set on which there is a mapping $d$ which measures the 'distance' between every pair of points in $V$.

The precise requirements on $d$ are that, for all $x, y$ and $z$ in $V$,

(i) $d(x, y) \geq 0$ and $d(x, y) = 0$ if and only if $x = y$,
(ii) $d(x, y) = d(y, x)$,
(iii) $d(x, z) \leq d(x, y) + d(y, z)$.

If these conditions are satisfied, then $d$ is called a *metric* on $V$ and the pair $(V, d)$ is called a *metric space.*

Since Devaney framed his definition of chaos, several authors have investigated the extent to which the conditions (a), (b) and (c) are independent of each other. In 1992 five Australian mathematicians Banks, Brooks, Cairns, Davis and Stacey obtained the surprising result that the condition (a) is a consequence of conditions (b) and (c) so that, in Devaney's definition, the condition (a) is redundant.

In the special case of symmetric one-hump mappings, we can prove much more: in the next section we prove the converse of each of the lemmas 9.1.1, 9.2.2 and 9.3.1. It follows that, for symmetric one-hump mappings, any one of the conditions (a), (b) or (c) implies the existence of wiggly iterates (provided the sensitivity constant $\delta \geq 1/2$). Thus for these mappings, any one of the conditions (a), (b) or (c) implies the other two.

## Exercises 9.4

9.4.1. Verify that $(V, d)$ is a metric space if $V = \mathbb{R}$ and $d(x, y) = |x - y|$. (This choice of $d$ is called the *usual metric* on $\mathbb{R}$.)

## 9.5 *CONVERSE LEMMAS

In this section the *converses* of Lemmas 9.1.1, 9.2.2, and 9.3.1 will be proved. Thus it will be shown that if a one-hump mapping has any one of

(a) *sensitive dependence,*
(b) *transitivity,*
(c) *a dense set of periodic points,*

then the mapping has wiggly iterates.

Each of the three proofs has a common logical structure, based on the idea of the contrapositive of an *if... then* statement.

> The *Contrapositive* of a statement of the form
> *if P then Q*
>
> is the statement
> *if not Q then not P.*
>
> Each *if... then* statement is logically equivalent to its contrapositive. To prove the original *if... then* statement, it therefore is valid to prove its contrapositive.

Hence in two of the proofs we start by supposing that the mapping does not have wiggly iterates. From this supposition we then derive a contradiction to one of the properties (a) or (c).

In the proof for property (b), we take a direct approach using the Intermediate Value Theorem and one of its consequences – the fact that under a continuous map, the image of an interval is an interval (see Exercise 9.5.1).

**9.5.1 Lemma (Sensitive implies wiggly)**
*Let $f : [0,1] \to [0,1]$ be a symmetric one–hump mapping. If $f$ has sensitive dependence with $\delta \geq \frac{1}{2}$, then $f$ has wiggly iterates.*

*Proof:* Suppose that $f$ does not have wiggly iterates, so by Lemma 7.3.2 there is an interval $K$ which does not contain zeroes of any $f^n$.

To show that $f$ does not have sensitive dependence (with $\delta \geq \frac{1}{2}$).

We shall show that the graph of $f^n$ must contain a hump like the ones shown in Figure 9.5.1.

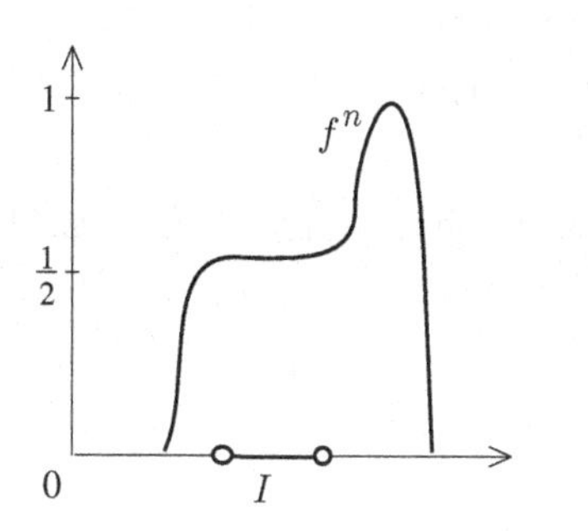


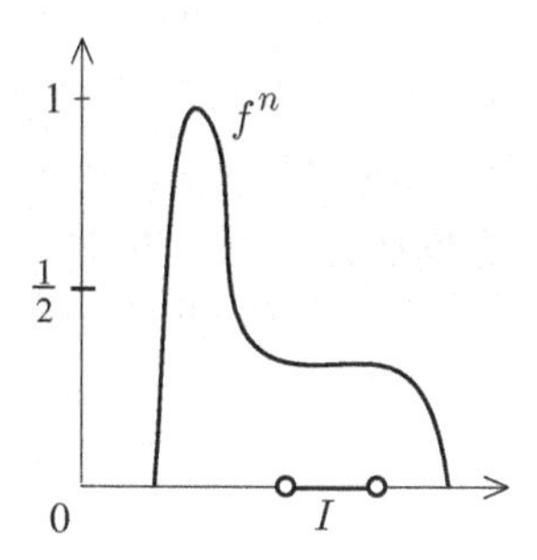


**Figure 9.5.1** Possible graphs of $f^n$

Since $f$ is symmetric, $\frac{1}{2}$ is a backward iterate of 0. Hence $f^n$ cannot assume the value $\frac{1}{2}$ at any point of $K$.

By the Intermediate Value Theorem, the values which $f^n$ assumes at points of the interval $K$, must therefore be

*all* bigger than $\frac{1}{2}$ or all less than $\frac{1}{2}$.

In both cases this implies that the values of $f^n$ must lie in an interval of length $\frac{1}{2}$. Hence for each $x, y \in K$,

$$|f^n(x) - f^n(y)| < \frac{1}{2}.$$

This means that $f$ cannot have sensitivity constant $\frac{1}{2}$, contrary to the hypothesis of the lemma. ■

**9.5.2 Lemma (Transitive implies wiggly)**
*Let $f : [0, 1] \to [0, 1]$ be a one–hump mapping. If $f$ is transitive then $f$ has wiggly iterates.*

*Proof:* Let $c$ denote the critical point of $f$.

By Lemma 7.3.2, it will suffice to show that every interval $K \subseteq [0, 1]$ contains a zero of some iterate of $f$.

Since $f$ is transitive, there is an $m$ such that $f^m(K) \cap K \neq \varnothing$. Hence there is an $x$ in $K$ such that $f^m(x) \in K$.

Since $f^m$ is continuous and $f^m(1) = 0$, there is an $r \in (c, 1)$ such that $f^m(J) \subseteq [0, c)$ where $J$ is the interval $(r, 1]$. (see Figure 9.5.2)

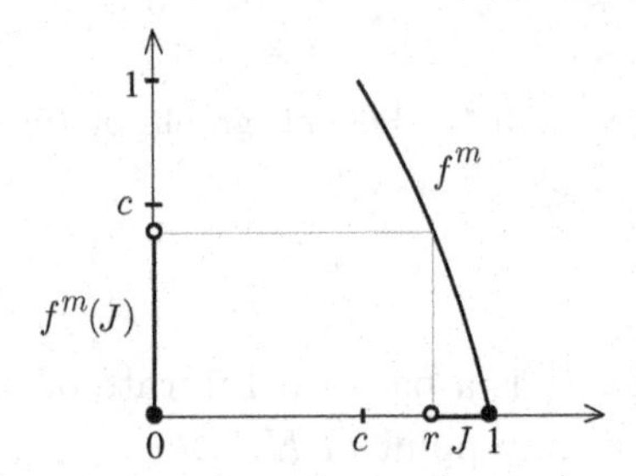


**Fig. 9.5.2** $f^m$ must map some interval $J = (r, 1]$ into $[0, c)$.

Transitivity of $f$ gives $n$ such that $f^n(K) \cap J \neq \varnothing$ and hence a $y$ in $K$ such that $f^n(y) \in J$. Hence

$$f^m(f^n(y)) \in f^m(J) \subseteq [0, c) \implies f^m(f^n(y)) = f^{m+n}(y) < c \qquad (1)$$

If $c \in f^n(K)$ we are done because $c$ is a zero of $f^2$ and hence $K$ contains a zero of $f^{n+2}$.

Otherwise, $f^n(K)$ is an interval by Exercise 9.5.1.

Since $f^n(K)$ intersects $J \subseteq (c, 1]$, we now deduce that $f^n(K) \subseteq (c, 1]$. Hence using the fact that $f^m(x) \in K$:

$$f^n(f^m(x)) \in f^n(K) \subseteq (c, 1] \implies f^n(f^m(x)) = f^{m+n}(x) > c \qquad (2)$$

Applying (1), (2) and the Intermediate Value Theorem to the continuous map $f^{m+n}$ yields a $z$ between $x$ and $y$ such that $f^{m+n}(z) = c$ (see Figure 9.5.3).

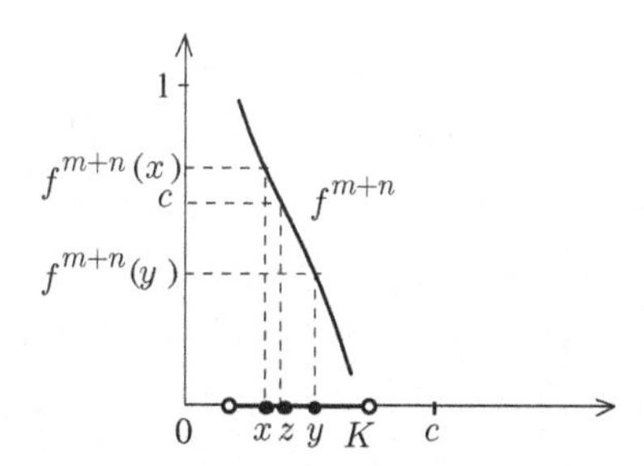


**Fig. 9.5.3** Applying the Intermediate Value Theorem.

Now $c$ is a zero of $f^2$, so $z$ is a zero of $f^{m+n+2}$.

Since $x, y \in K$ and $z$ is between $x$ and $y$, $z \in K$. ∎

The next lemma will be used in the proof of our final converse lemma which shows that a one hump map with a dense set of periodic points must have wiggly iterates.

**9.5.3 Lemma** *Let $f : [0,1] \to [0,1]$ be a one-hump mapping. Suppose $J$ is an interval containing no zeroes of any $f^n$. Then there is an interval $K$ containing no zeroes of any $f^n$ such that $f^n$ is strictly decreasing on at least one of the intervals $J$ and $K$ for all $n \geq 1$.*

*Proof:* In fact we will prove a slightly stronger statement:

(*) Each $f^n$ is strictly increasing on one of the intervals $J$ and $K$ and strictly decreasing on the other.

Let $c$ be the critical point of $f$.

Exercise 9.5.2 gives an interval $K$ such that $f(J) = f(K)$ and either $f$ is strictly increasing on $J$ and strictly decreasing on $K$ or vice-versa.

This shows that (*) is true for $n = 1$. We will complete the proof by induction. Suppose (*) is true for some $n \geq 1$.

We will assume that $f^n$ is strictly increasing on $J$ and strictly decreasing on $K$. The proof of the other case is similar.

Since $f(J) = f(K)$, we have $I = f^n(J) = f^n(K)$ and since $J$ contains no zeroes of iterates of $f$, neither does $I$.

In particular, $c \notin I$, so either $I \subseteq [0, c)$ or $I \subseteq (c, 1]$.

We take these two cases separately.

*Case (i)*: If $I \subseteq [0, c)$, then $f$ is strictly increasing on $I$ so by Exercise 9.5.4, $f^{n+1}$ is strictly increasing on $J$ and strictly decreasing on $K$.

*Case (ii)*: If $I \subseteq (c, 1]$, then $f$ is strictly decreasing on $I$ so by Exercise 9.5.4, $f^{n+1}$ is strictly decreasing on $J$ and strictly increasing on $K$.

In either case, (*) is true for $n + 1$ and the lemma holds by mathematical induction. ∎

**9.5.4 Lemma (Dense periodic points implies wiggly)**
*Let $f : [0, 1] \to [0, 1]$ be a one-hump mapping. If $f$ has a set of periodic points which is dense in $[0, 1]$, then $f$ has wiggly iterates.*

*Proof:* We prove the contrapositive of the statement in the theorem.

Suppose that $f$ does not have wiggly iterates, so by Lemma 7.3.2, there is then an interval $J$ which contains no zeroes of any $f^n$.

Let $K$ be the interval given in the statement of Lemma 9.5.3.

To prove the contrapositive of the statement in the Lemma, it remains to show that the periodic points are not dense in $[0, 1]$. We prove this by contradiction.

Suppose that the periodic points are dense in $[0, 1]$.

Hence they are dense in the interval $J \subseteq [0, 1]$.

Thus there are two different periodic points in $J$, say $p_1 < p_2$ with periods say $m_1$ and $m_2$.

Likewise there are points $q_1 < q_2$ in $K$ with periods $n_1$ and $n_2$.

Put $n = m_1 m_2 n_1 n_2$ so that $f^n$ has $p_1, p_2, q_1$ and $q_2$ as fixed points.

Lemma 9.5.3 tells us that $f^n$ is strictly decreasing on $J$ or $K$.

If $f^n$ is decreasing on $J$, we have $f^n(p_1) = p_1 > p_2 = f^n(p_2)$.

If $f^n$ is decreasing on $K$, we have $f^n(q_1) = q_1 > q_2 = f^n(q_2)$.

In either case we have a contradiction. We conclude that the periodic points of $f$ are not dense in $[0, 1]$. ∎

## Exercises 9.5

9.5.1. First satisfy yourself that the following formal definition of an interval in $\mathbb{R}$ works by considering its application to the various types of intervals (open, closed, half-open, unbounded etc).

> A set $I \subseteq \mathbb{R}$ is an *interval* if for every $x, y \in I$, every real number between $x$ and $y$ is also in $I$.

Now use this definition together with the Intermediate Value Theorem to prove that if $f : \mathbb{R} \to \mathbb{R}$ is continuous and $I$ is an interval, then $f(I)$ is also an interval.

9.5.2. Let $f : [0, 1] \to [0, 1]$ be a one hump mapping with critical point $c$ and let $J \subseteq [0, 1]$ be an interval containing no zeroes of any $f^n$.

(a) Show $f$ is either strictly increasing or strictly decreasing on $J$.

(b) Find another interval $K$ such that $f(J) = f(K)$ and such that if $f$ is strictly increasing on $J$, then $f$ is strictly decreasing on $K$ and vice versa.

[Hint: Draw a picture. $J$ and $K$ must be on opposite sides of $c$. You will need the Intermediate Value Theorem.]

9.5.3. Let $f : [0, 1] \to [0, 1]$ be a one hump mapping and let $c$ be the critical point.

(a) Use the Intermediate Value Theorem to prove that $f$ has a fixed point $p$ in $(c, 1)$.

(b) Modify the proof of Lemma 9.5.2 to show that the set of backward iterates of $p$ is dense.

[Hint: Let $p$ play the role that $c$ played in Lemma 9.5.2.]

9.5.4. Prove the the following:

(a) A composite of two strictly increasing maps is strictly increasing.

(b) A composite of two strictly decreasing maps is strictly increasing.

(c) A composite of a strictly increasing map and a strictly decreasing map is strictly decreasing.

---

## Additional reading for Chapter 9

There are many books discussing the consequences of chaos for predictive sciences. Some are listed in the bibliography below.

The proof that dense periodic points together with transitivity imply Sensitive dependence is in [BBCDS]. This was also proved independently by Stephen Silverman in [Sil]. An expanded version of the proof is to be found in Chapter 9 of [Ho].

An animated explanation of the idea of chaos, focussing mainly on discrete models can be found on the World Wide Web at

http://johnbanks.maths.latrobe.edu.au/chaos/.

---

## References

[AP] D.K. Arrowsmith and C.M. Place, *Dynamical Systems: Differential Equations, Maps and Chaotic Behaviour,* Chapman and Hall, 1992.

[Ba] Michael F. Barnsley, *Fractals Everywhere,* 2nd edition, Academic Press Professional, Boston 1993.

[BBCDS] John Banks, Jeff Brooks, Grant Cairns, Gary Davis and Peter Stacey, "On Devaney's Definition of Chaos", *American Mathematical Monthly,* **99** (1992), 332-334.

[BG] G. Baker and J. Gollub, *Chaotic Dynamics – an Introduction,* Springer-Verlag, 1992.

[De1] Robert L. Devaney, *An Introduction to Chaotic Dynamical Systems: Second Edition,* Addison-Wesley, Menlo Park California, 1989.

[Gl] James Gleick, *Chaos: Making a New Science,* Sphere Books, London, 1988.

[Ho] Richard Holmgren, *A First Course in Discrete Dynamical Systems* Springer-Verlag, 1994.

[PJS] Hans-Otto Peitgen, Hartmut Jürgens and Dietmar Saupe, *Chaos and Fractals, New Frontiers of Science*, Springer-Verlag, New York, 1992.

[Ru1] David Ruelle, *Elements of Differentiable Dynamics and Bifurcation Theory* Academic Press 1989.

[Ru2] David Ruelle, *Chance and Chaos,* Penguin 1991.

[Sil] Stephen Silverman, "On Maps with Dense Orbits and the Definition of Chaos", *Rocky Mountain Journal of Mathematics,* **22**(1992), 353-375.

[Ste] Ian Stewart, *Does God Play Dice?* Penguin Books, Middlesex, 1989.

# 10

# WIGGLES, 'WOGGLES' AND SCHWARZIAN DERIVATIVES

*The aim of this chapter is to provide an easy-to-use test for showing that a mapping* $f : [0,1] \to [0,1]$ *has chaotic behaviour. The test is applicable to symmetric one-hump mappings which are differentiable three times and involves the concept of the Schwarzian derivative of* $f$.

*In earlier chapters we approached the idea of chaotic behaviour with the aid of the concept of wiggly iterates. Hence we begin this chapter by investigating the behaviour of some mappings which do not have wiggly iterates. This leads to the concept of a* fat wiggle *or a* woggle.

*Schwarzian derivatives are then introduced as a device to test for the absence of woggles.*

## 10.1 WIGGLES AND WOGGLES

If $f : [0,1] \to [0,1]$ is a one-hump mapping, then its $n$th iterate will consist of $2^{n-1}$ humps. Each hump has the interval between two consecutive zeroes of $f^n$ as its base. As $n \to \infty$ the number of wiggles in the $n$th iterate also approaches infinity.

How can such a mapping fail to have wiggly iterates? The answer to this question is suggested by looking at the computer plots of some examples.

### Failure to wiggle

Graphs of two examples of one-hump mappings which do not have wiggly iterates are shown in Figure 10.1.1. Each mapping was obtained by modifying the formula for the logistic mapping $Q_4$.
The first mapping is $f : [0,1] \to [0,1]$ with

$$f(x) = 16x^2(1-x)^2. \tag{1}$$

The second mapping is $g : [0,1] \to [0,1]$ with

$$g(x) = \frac{4x(1-x)}{1 + 1000x^2(x-0.5)^2(x-1)^2}. \tag{2}$$

The mapping $f$ has an attracting fixed point at 0, while the mapping $g$ has a peculiar shape on which we shall later base our definition of a *woggle*.

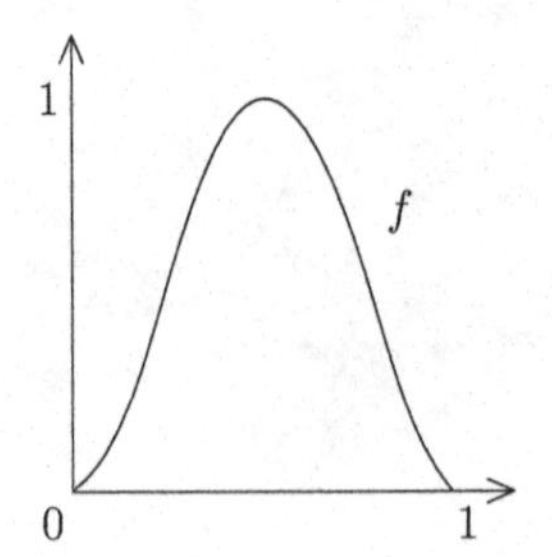


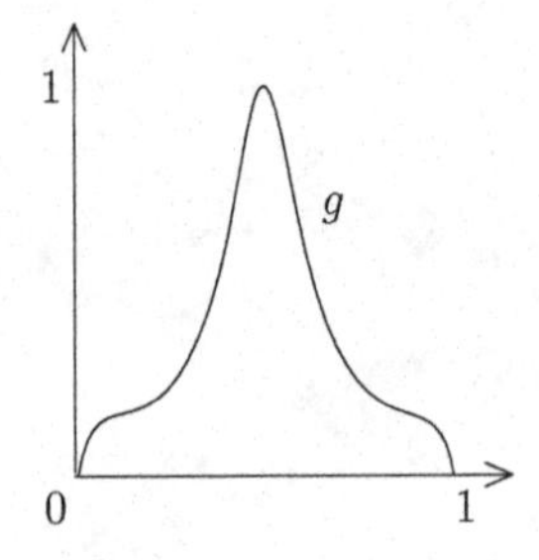


**Figure 10.1.1** Two mappings lacking wiggly iterates.

Graphs of the next three iterates of $f$ and $g$ are shown in Figure 10.1.2. For both mappings, the number of humps doubles at each iteration. The graphs suggest that, instead of the bases of the humps all getting small, the bases of some wiggles appear to remain large, forcing the bases of the other wiggles to be correspondingly small. This behaviour of the iterates is in marked contrast to that observed in Section 7.2 for the iterates of the logistic mapping $Q_4$ and of the tent mapping $T_4$.

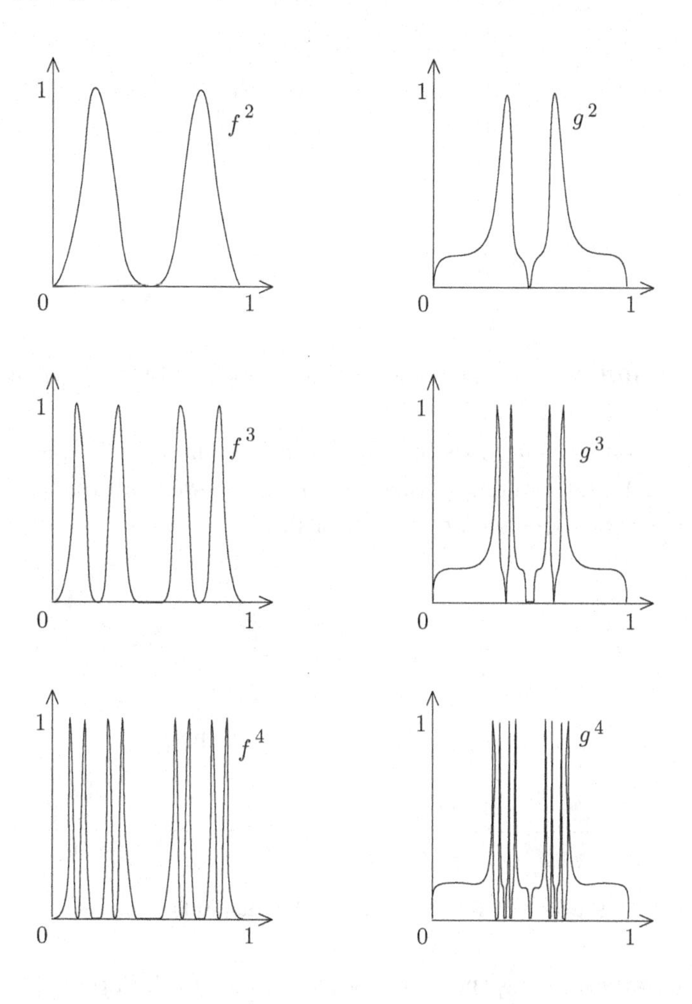


**Figure 10.1.2.** Iterates of the mappings $f$ and $g$.

## Woggles

The computer plots suggest that a one-hump mapping can fail to have wiggly iterates because the maximum length of the base of a hump need not approach 0 as the number of iterations approaches infinity. Some humps, or wiggles, stay fat while their neighbours get thinner and thinner. A hypothetical case is shown in Figure 10.1.3.

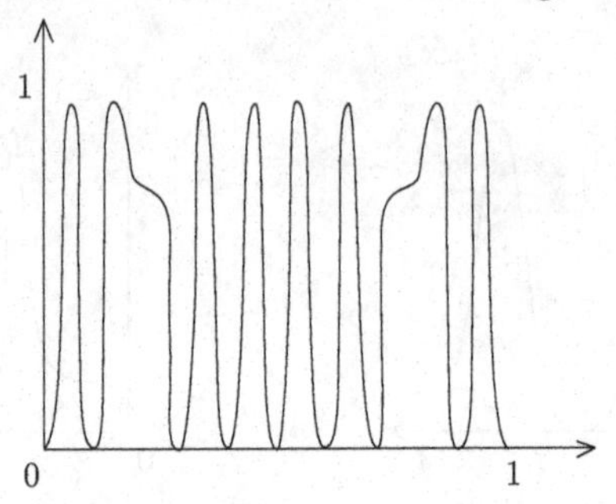


**Figure 10.1.3** An iterate of a one-hump mapping without wiggly iterates.

These 'fat' wiggles often have a distinctive shape, which is analyzed in Figure 10.1.4. To emphasize the difference, we shall refer to humps or wiggles with this shape as 'woggles'.

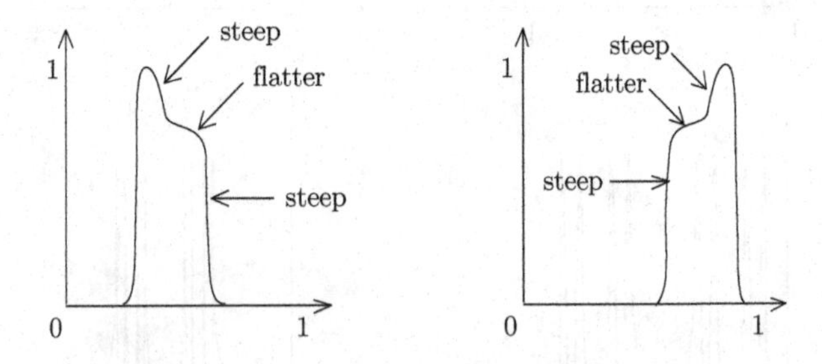


**Figure 10.1.4.** Some possible shapes for woggles.

Thus we are using the word *woggle* to mean a hump of a mapping whose base contains an interval $I$ on which the slope of the hump

(a) is either always positive or always negative and

(b) follows the pattern

$$\textit{steep} \quad \longrightarrow \quad \textit{flatter} \quad \longrightarrow \quad \textit{steeper}$$

### Woggles and derivatives

We now assume that the mapping $f$ is differentiable and that its derivative $f'$ is continuous (which is usually expressed by saying that $f$ is $\mathcal{C}^1$; more generally, saying that $f$ is of class $\mathcal{C}^n$ is just a shorthand for saying that its first $n$ derivatives exist and are continuous). We aim to express the idea that $f$ has a woggle in terms of the behaviour of $f'$. We distinguish between two cases.

*First Case:* the woggle occurs on a portion of the hump where the mapping $f' > 0$, as in the graph on the right of Figure 10.1.4. In this case, the slope $f'$ follows the development

*large positive* $\longrightarrow$ *smaller positive* $\longrightarrow$ *large positive.*

Hence the slope $f'$ has a *positive local minimum* at some point $a \in I$.

*Second Case:* the woggle occurs on a portion of the hump where the mapping $f' < 0$ , as in the graph on the left of Figure 10.1.4. In this case, the slope $f'$ follows the development

*large negative* $\longrightarrow$ *smaller negative* $\longrightarrow$ *large negative.*

Hence the slope $f'$ has a *negative local maximum* at some point $a \in I$.

The two cases may be combined to give the following precise definition of a woggle.

**10.1.1 Definition** A *woggle* is a hump of a $\mathcal{C}^1$ mapping $f$ whose base contains an interval $I$ and a point $a \in I$ such that

(a) $f' > 0$ on $I$ and $f'$ has a positive local minimum at $a$, or

(b) $f' < 0$ on $I$ and $f'$ has a negative local maximum at $a$. ∎

We now use a well-known test from calculus for maxima and minima to derive a necessary condition for the existence of a woggle.

If a $C^2$ function $\phi$ has a *local minimum* at a point $a$ of an open interval $I$ then

$$\phi'(a) = 0 \qquad \text{and} \qquad \phi''(a) \geq 0$$

and, similarly, if it has a *local maximum* at $a$ then

$$\phi'(a) = 0 \qquad \text{and} \qquad \phi''(a) \leq 0.$$

Applying this result with the choice $\phi = f'$ and using the definition of a woggle we deduce that *if a $C^3$ mapping $f$ has a woggle on an interval $I$ then there is a point $a \in I$ such that*

$$f' > 0 \ \textit{on}\ I \qquad \textit{and} \qquad f''(a) = 0, \quad f'''(a) \geq 0; \tag{3}$$

*or*

$$f' < 0 \ \textit{on}\ I \qquad \textit{and} \qquad f''(a) = 0, \quad f'''(a) \leq 0. \tag{4}$$

By combining (3) and (4) we get the following lemma.

**10.1.2 Lemma** *If $f$ has a woggle there is an open interval $I$ contained in the base of the hump, such that*

*(a) $f' > 0$ on $I$ or $f' < 0$ on $I$ and*
*(b) there is a point $a \in I$ such that*

$$f'(a)f'''(a) \geq 0 \qquad \text{and} \qquad f''(a) = 0. \tag{5}$$

## Exercises 10.1

10.1.1. How many woggles appear in Figure 10.1.2 of the text
(a) in the graph of $g^2$?
(b) in the graph of $g^3$?

10.1.2. Give an example of a mapping $f : \mathbb{R} \to \mathbb{R}$ such that $f$ is strictly monotonic on $\mathbb{R}$ but $f'(0) = 0$.

10.1.3.

(a) Explain why a mapping which has an attracting fixed point cannot have wiggly iterates.

(b) Prove that the mapping $f$ given by (1) in the text does not have wiggly iterates.

10.1.4. The relationship between the graph of id and the graph of the mapping $g$ given by (2) in the text is shown below in Figure 10.1.5. Prove that the mapping $g$ cannot have wiggly iterates.

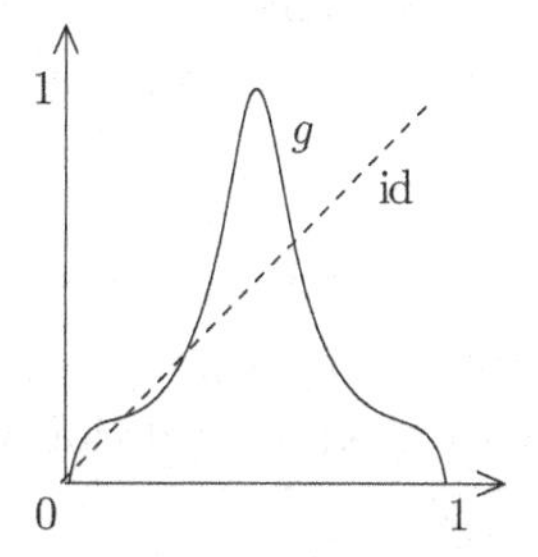


**Figure 10.1.5.**

10.1.5. Prove that either of the conditions (3) and (4) in the text implies the condition (5).

10.1.6. Let $f$ be a function with the property that there is an open interval $I$ in its domain such that $f' > 0$ on $I$ and $f'$ has a *positive local maximum* at some point $a \in I$.

(a) Sketch the graph of a function which illustrates this property.

(b) Show that $f'(a)f'''(a) \leq 0$ and $f''(a) = 0$.

10.1.7. Let $f$ be a function with the property that there is an open interval $I$ in its domain such that $f' < 0$ on $I$ and $f'$ has a *negative local minimum* at some point $a \in I$.

(a) Sketch the graph of a function which illustrates this property.

(b) Show that $f'(a)f'''(a) \leq 0$ and $f''(a) = 0$.

10.1.8. If a one-hump mapping is differentiable, what is the least number of zeroes which the derivative of its $n$th iterate can have?

10.1.9. *A character building exercise.* Prove that the mapping $g$ given by equation (2) in the text is a one-hump mapping.

## 10.2 THE SCHWARZIAN DERIVATIVE

As shown in the previous section, one way in which a mapping may fail to have wiggly iterates is by having a woggle. In this section we devise a simple test to guarantee the absence of woggles in a mapping and its iterates.

### Avoiding woggles

Lemma 10.1.2 shows that the mapping $f$ *cannot* have a woggle on an interval $I$ if the following condition is satisfied at every point $a \in I$:

$$f'(a)f'''(a) < 0 \quad \textit{whenever} \quad f''(a) = 0. \tag{6}$$

We use this condition to motivate the idea of the Schwarzian derivative. Since we want to avoid woggles, not only in the original mapping but also in its iterates, we need a condition which is preserved under composition.

To provide some flexibility, we therefore introduce a constant $c > 0$ and replace (6) by the condition

$$f'(a)f'''(a) < c(f''(a))^2 \quad \textit{whenever} \quad f''(a) = 0. \tag{7}$$

The condition (7) also is sufficient to prevent woggles appearing in the graph of $f$. It is also sufficient to have

$$f'f''' - c(f'')^2 < 0.$$

Thus we have proved the following result.

**10.2.1 Lemma** *Let $f : [0,1] \to [0,1]$ be a three times differentiable symmetric one-hump mapping. Then $f$ cannot have a woggle on an interval $I$ if*

$$f'(x)f'''(x) - c(f''(x))^2 < 0. \tag{8}$$

*at all points $x \in I$ at which $f'(x) \neq 0$*

This lemma is valid no matter how we choose the constant $c > 0$. It turns out, however, there is one particular choice, $c = \frac{3}{2}$, which miraculously ensures that the property (8) is preserved under composition. To avoid fractions, it is convenient to multiply (8) through by 2. This is our motivation for the following definition.

**10.2.2 Definition** Let $f : I \to I$ be three times differentiable. We define the *Schwarzian derivative*[1] of $f$ by putting

$$S(f)(x) = 2f'(x) \cdot f'''(x) - 3(f''(x))^2$$

for all $x \in I$ such that $f'(x) \neq 0$. ■

Note that the Schwarzian derivative $S(f)$ can have a smaller domain than $f$ itself because we omit the points at which $f' = 0$. We say that $f$ has *negative Schwarzian derivative* if the condition $S(f) < 0$ holds at all points of the domain of $S(f)$; that is, at all points for which $f' \neq 0$.

**Invariance under composition.**

We now prove a lemma which is the key to showing that negative Schwarzian derivatives are preserved under composition.

**10.2.3 Lemma** $S(f \circ g) = S(f) \circ g \cdot (g')^4 + (f' \circ g)^2 \cdot S(g)$.

*Proof:* Note that, by the chain rule,

$$\begin{aligned}
(f \circ g)' &= f' \circ g \cdot g' \\
(f \circ g)'' &= f'' \circ g \cdot (g')^2 + f' \circ g \cdot g'' \\
(f \circ g)''' &= f''' \circ g \cdot (g')^3 + 3f'' \circ g \cdot g' \cdot g'' + f' \circ g \cdot g'''
\end{aligned}$$

Hence, by definition of the Schwarzian derivative, $S(f \circ g) = A + B$, where

$$\begin{aligned}
A &= 2(f \circ g)' \cdot (f \circ g)''' \\
&= 2f' \circ g \cdot g' \cdot (f''' \circ g \cdot (g')^3 + 3f'' \circ g \cdot g' \cdot g'' + f' \circ g \cdot g''') \\
&= 2f''' \circ g \cdot f' \circ g \cdot (g')^4 + 6f'' \circ g \cdot f' \circ g \cdot g'' \cdot (g')^2 + 2(f' \circ g)^2 \cdot g''' \cdot g'
\end{aligned}$$

and

$$\begin{aligned}
B &= -3((f \circ g)'')^2 \\
&= -3(f'' \circ g \cdot (g')^2 + f' \circ g \cdot g'')^2 \\
&= -3(f'' \circ g)^2 \cdot (g')^4 - 6f'' \circ g \cdot f' \circ g \cdot g'' \cdot (g')^2 - 3(f' \circ g)^2 \cdot (g'')^2.
\end{aligned}$$

[1] Our definition of the Schwarzian derivative differs somewhat from the standard definition, but is in line with the approach in [Mil]. See Exercise 10.2.11 for a comparison with the standard definition.

Hence

$$\begin{aligned} S(f \circ g) &= A + B \\ &= 2f''' \circ g \cdot f' \circ g \cdot (g')^4 + 2(f' \circ g)^2 \cdot g''' \cdot g' \\ &\qquad - 3(f'' \circ g)^2 \cdot (g')^4 - 3(f' \circ g)^2 \cdot (g'')^2 \\ &= S(f) \circ g \cdot (g')^4 + (f' \circ g)^2 \cdot S(g). \end{aligned}$$

■

**10.2.4 Theorem** *If $f$ and $g$ both have negative Schwarzian derivative, then so does their composite $f \circ g$.*

*Proof:* Suppose that both $f$ and $g$ have negative Schwarzian derivatives.

We show that $f \circ g$ has negative Schwarzian derivative.
By Definition 10.2.2 this means showing that

$$S(f \circ g)(x) < 0$$

for all $x$ such that $(f \circ g)'(x) \neq 0$.

Let $x$ be such that $(f \circ g)'(x) \neq 0$;
that is $f'(g(x)) \neq 0$ and $g'(x) \neq 0$, by the chain rule.

Hence Lemma 10.2.3 gives

$$S(f \circ g) = S(f) \circ g \cdot (g')^4 + (f' \circ g)^2 \cdot S(g).$$

where, at the relevant values of $x$, the factors $(g')^4$ and $(f' \circ g)^2$ are positive. Hence

$$S(f \circ g) < 0.$$

■

An important consequence of this lemma is that if a mapping has negative Schwarzian derivative, then so do all of its iterates.

## Exercises 10.2

10.2.1.

(a) Show that $f$ has negative Schwarzian derivative where $f(x) = x^2$.

(b) Show that $f$ has negative Schwarzian derivative where $f(x) = ax^2 + bx + c$ with $a \neq 0$.

10.2.2. What is the Schwarzian derivative of the tent mapping $T_4$?

10.2.3. Sketch the graph of the cubing function $f : \mathbb{R} \to \mathbb{R}$ with $f(x) = x^3$. Show that $f$ has negative Schwarzian derivative at each $x$ for which $f'(x) \neq 0$.

10.2.4. If we had defined the Schwarzian derivative of $f$ by putting

$$S(f) = f' \cdot f''' - c(f'')^2$$

with $c = -2$, what change would result in Lemma 10.2.3 ? Would Theorem 10.2.4 still be valid?

10.2.5. Show that if $f$ has negative Schwarzian derivative then so does the mapping $-f$.

10.2.6. Show that if $f$ has negative Schwarzian derivative then so do $f + c$ and $cf$ for each real number $c \neq 0$.

10.2.7. Verify that the property of having zero Schwarzian derivative is preserved under composition of mappings.

[Hint. Use Lemma 10.2.3.]

10.2.8. Verify that the property of having positive Schwarzian derivative is preserved under composition of mappings.

[Hint. Use Lemma 10.2.3.]

10.2.9. Suppose that a mapping $f$ has *positive* Schwarzian derivative. What type of behaviour does this prevent occurring on the graph of $f$?

[Hint. See Exercises 10.1.6 and 10.1.7.]

10.2.10. Suppose that $f : [0,1] \to [0,1]$ is a symmetric one-hump mapping with negative Schwarzian derivative.

(a) Give a graphical argument to show that $f$ has exactly two fixed points provided $f'(0) > 1$.

(b) Show that $f(x) > x$ for all $x \in [0, \frac{1}{2}]$ provided $f'(0) > 1$.

(c) Show that the multiplier of the nonzero fixed point must be less than $-1$.

[Hint. Suppose not and then derive a contradiction by considering the graph of $f^2$ and showing it must have a woggle.]

10.2.11. The standard definition of the Schwarzian derivative of a map $f$ at a point $x$ such that $f'(x) \neq 0$ is given by

$$\overline{S}(f)(x) = \frac{f'''(x)}{f'(x)} - \frac{3}{2}\left(\frac{f''(x)}{f'(x)}\right)^2.$$

Show that $\overline{S}(f)(x)$ is negative if and only if $S(f)(x)$ is negative.

## 10.3 TESTING FOR CHAOS

Two examples of mappings which failed to have wiggly iterates were graphed in Figure 10.1.1. One of these mappings, $f$, had an *attracting fixed point at the origin* while the other, $g$, had *a woggle.*

The following theorem shows that these are the only two ways in which a symmetric one-hump mapping can fail to have wiggly iterates.

**10.3.1 Theorem** *Let $f$ be a symmetric one-hump mapping. If $f$ fails to have wiggly iterates then $f'(0) \leq 1$ or some $f^n$ has a woggle.*

The proof of this theorem, which is rather long, is deferred until Section 10.4. Meanwhile, we deduce the following corollary and then show how this corollary can be used to test a mapping for chaotic behaviour.

**10.3.2 Theorem (Test for chaos)**

*Let $f$ be a symmetric one-hump mapping. If $f'(0) > 1$ and if $f$ has negative Schwarzian derivative (except at $x = \frac{1}{2}$), then $f$ has chaotic behaviour.*

*Proof:* Let $f$ be a symmetric one-hump mapping.

Suppose that $f'(0) > 1$ and that $f$ has negative Schwarzian derivative.

> In Theorem 10.3.1, the *if... then* statement is logically equivalent to its contrapositive, which is:
>
> *If $f'(0) > 1$ and all the $f^n$'s have no woggles, then $f$ has wiggly iterates.*

By the argument of Section 10.2, each $f^n$ is free of woggles.

It therefore follows from the contrapositive form of Theorem 10.3.1 that $f$ has wiggly iterates.

Hence $f$ is chaotic. ∎

Recall that, in Section 7.2, we showed the mapping $Q_4$ had wiggly iterates by using the formula for the $n$th iterate of $x_0$. The following example shows how much easier it is to establish the chaotic behaviour of $Q_4$ by using Theorem 10.3.2.

**10.3.3 Example** *Use the test to prove that the logistic mapping $Q_4$ has chaotic behaviour.*

*Solution:* We apply Theorem 10.3.2 to $f : [0, 1] \to [0, 1]$ where

$$f(x) = 4x(1 - x).$$

Hence $f'(0) = 4 > 1$ while the Schwarzian derivative of $f$ is given by

$$\begin{aligned} S(f)(x) &= 2f'(x)f'''(x) - 3(f''(x))^2 \\ &= 0 - 3 \times (-8)^2 = -192 < 0. \end{aligned}$$

Thus $f$ has negative Schwarzian derivative.

From Theorem 10.3.2, $f$ is chaotic. ∎

This example was particularly easy because the Schwarzian derivative turned out to be constant, making it easy to determine whether the Schwarzian derivative was negative. This isn't always the case, as the following exercises show.

## Exercises 10.3

10.3.1. Show that the following mappings $f : [0, 1] \to [0, 1]$ are chaotic

$$f(x) = 1 - (2x - 1)^4$$
$$f(x) = \sin(\pi x)$$
$$f(x) = 4x(1 - x)e^{3(x - \frac{1}{2})^2}.$$

## 10.4 *PROVING THEOREM 10.3.1

One of the main tools we need to prove Theorem 10.3.1 is the *Mean Value Theorem* first introduced in Section 8.3, which may be restated as saying that if $g$ is differentiable on the closed interval $[a, b]$, then for some $x \in (a, b)$

$$g'(x) = \frac{g(b) - g(a)}{b - a}.$$

Geometrically, the Theorem means there is a point where the derivative of $g$ is equal to the gradient of the chord touching the graph of $g$ at the endpoints of the interval, as shown in Figure 10.4.1.

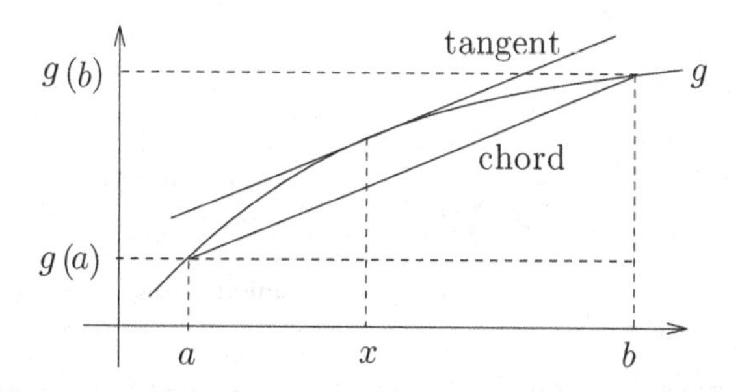


**Fig. 10.4.1** There is at least one $x$ between $a$ and $b$ where the chord and function have the same gradient.

Another well known result we use is the fact that if $g$ is continuous on an interval $[a, b]$,then there is a $k \geq 0$ such that $|g(x)| < k$ for all $x \in [a, b]$ (i.e., the mapping $g$ is *bounded* on $[a, b]$).

*Proof of Theorem 10.3.1:* Let $f$ be a symmetric one-hump mapping. Suppose that $f'(0) > 1$ and that $f$ does not have wiggly iterates.

We show some $f^n$ contains a woggle by first showing that the graph of some $f^n$ passes through what we shall call a 'pigeonhole'. We will break the proof of this first part of the proof into six steps.

**Step 1:** *There is a maximal open interval $I$ which contains no zeroes of iterates of $f$.*

*Proof:* By assumption, $f$ does not have wiggly iterates.

Hence the zeroes of the iterates of $f$ are not dense in $[0,1]$, by Lemma 7.3.2.

Thus there is an open interval $I_0 \subseteq [0,1]$ which contains no zeroes.

Take $I$ to be the union of all the open intervals in $[0,1]$ which

(i) contain the interval $I_0$ and

(ii) contain no zeroes of any iterates of $f$.

This choice of $I$ is maximal in the sense that there is no larger open interval satisfying (i) and (ii).

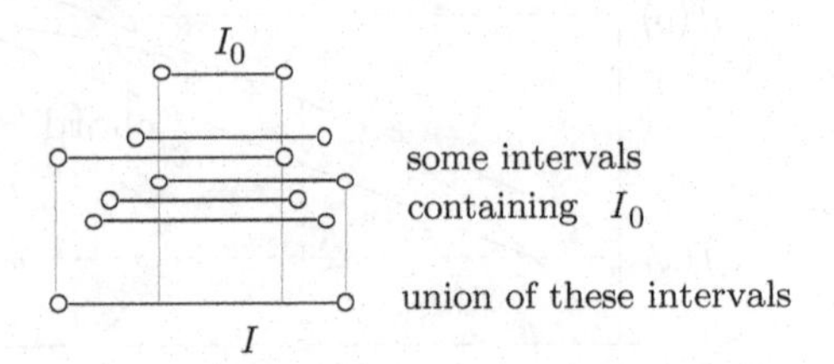


**Figure 10.4.2.** The union of some intervals containing $I_0$.

**Step 2:** *As close as we please to the endpoints of the interval $I$, there are consecutive zeroes $y$ and $z$ of some $f^n$.*

*Proof:* From (a), $I$ is a maximal open interval containing no zeroes of any iterate of $f$.

Hence there are zeroes $z_0$ and $z_1$ as close as we like to the endpoints of $I$.

But zeroes are preserved under iteration.

Hence we may assume that $z_0$ and $z_1$ are zeroes of the same $f^n$.

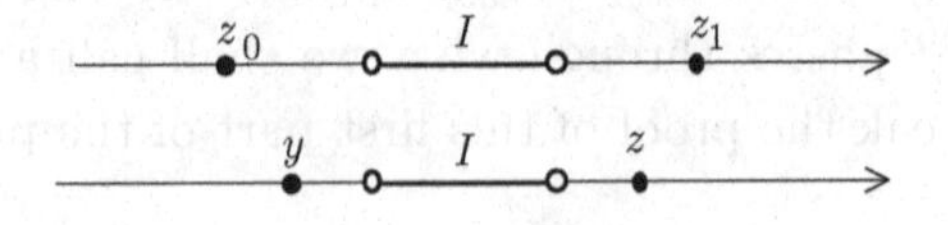


**Figure 10.4.3.** Zeroes as close as we like to either endpoint of $I$.

But since there are only finitely many zeroes of $f^n$, there must be a last (ie largest) one to the left of $I$, say $y$.

We take $z$ to be the first (ie smallest) zero on the right of $I$.

Both of these zeroes must be at least as close to $I$ as $z_0$ and $z_1$.

**Step 3:** *There is a decreasing sequence* $\{w_i\}_{i=1}^{\infty}$ *of zeroes with* $w_1 = \frac{1}{2}$, $f(w_{i+1}) = w_i$ *and* $\lim_{n\to\infty} w_i = 0$.

*Proof:* By Exercise 10.2.10(b), $f$ satisfies the hypotheses of Lemma 7.4.2. Setting $w_i = z_{i+1}$ in that Lemma gives the result.

**Step 4:** *There is a constant* $k > 1$ *such that* $w_i < kw_{i+1}$ *for all* $i$;

*Proof:* Since $f'$ is continuous there is a $k$ such that $|f'(x)| < k$ for all $x \in [0, 1]$. We may assume that $k > 1$.

Suppose that $f(x) \geq kx$ for some $x \in [0, 1]$. Then the Mean Value Theorem gives $z \in [0, x]$ such that

$$f'(z) = \frac{f(x) - f(0)}{x - 0} = \frac{f(x)}{x} \geq \frac{kx}{x} = k.$$

But this contradicts our assumption that $f'(x) < k$ for all $x \in [0, 1]$.

We conclude that $f(x) < kx$ for all $x \in [0, 1]$ and hence

$$w_i = f(w_{i+1}) < kw_{i+1}.$$

**Step 5:** *For each* $w_{i+1}$

$$\frac{k-1}{b-a} w_{i+1} < \frac{w_{i+1}}{a-y}, \frac{w_{i+1}}{z-b}, \frac{\frac{1}{2}}{z-b}, \frac{\frac{1}{2}}{a-y}.$$

*Proof:* Since $k - 1 > 0$, Step 2 lets us to choose $n$ large enough to get $y$ and $z$ sufficiently close to $I$ to ensure that

$$a - y < \frac{b-a}{k-1} \quad \text{and} \quad z - b < \frac{b-a}{k-1}.$$

Exercise 10.4.1 now yields the four required inequalities.

**Step 6:** *The graph of $f^n$ passes through a rectangle (a 'pigeon-hole') of the form*

$$I \times (w_{i+1}, w_i) \qquad \textit{or} \qquad I \times (1 - w_i, 1 - w_{i+1}).$$

*Proof:* Since $f$ is symmetric $f(1 - w_{i+1}) = f(w_{i+1}) = w_i$ and hence each $1 - w_i$ is a zero of $f$.

Thus $f^n$ cannot take any of the $w_{i+1}$ or $1 - w_{i+1}$ values (for any $i$) at any point in $I$, as this would contradict the fact that $I$ contains no zeroes.

Hence the graph of $f^n$ cannot cross the horizontal lines at

$$\cdots w_{i+1} < w_i < \cdots < w_1 = \frac{1}{2} < \cdots < 1 - w_i < 1 - w_{i+1} < \cdots$$

as shown in Figure 10.4.4. The graph must therefore pass through one of the 'pigeon-holes' bounded by these lines and the vertical lines at $a$ and $b$.

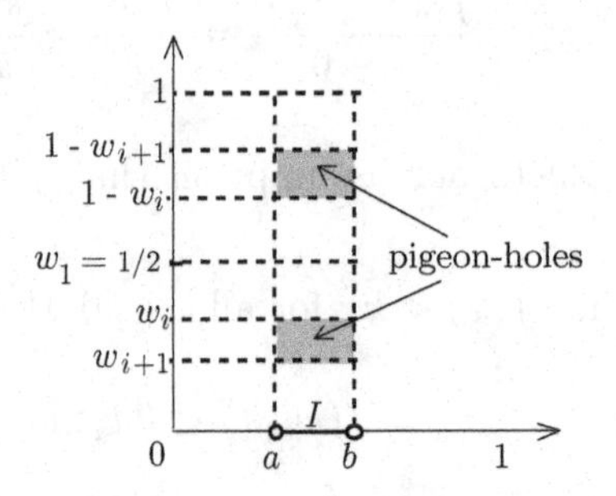


**Figure 10.4.4.** Pigeon-holes defined by the $w_i$.

As the graph passes through the pigeon-hole, the graph of $f^n$ may be increasing or decreasing. Since $f^n$ has a hump on $[y, z]$, there must be an $m \in [y, z]$ such that $f^n(m) = 1$. The point $m$ will be to the left of $I$ if $f^n$ is increasing on $I$ and to the right of $I$ if $f^n$ is decreasing on $I$. Figure 10.4.5 illustrates the two increasing cases.

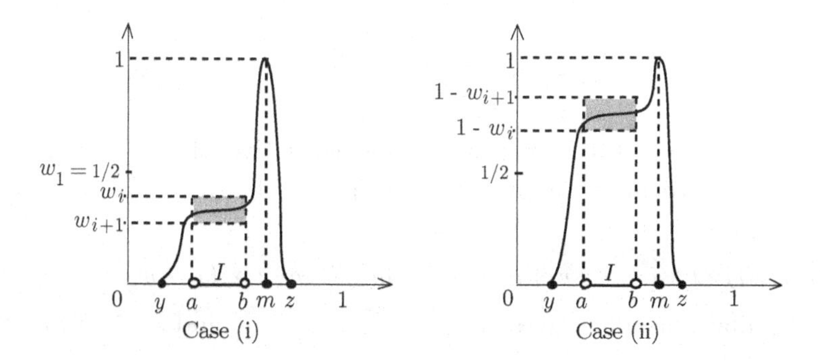


**Figure 10.4.5.** Constraints on the graph of $f^n$ in the case where $f^n$ is increasing on $I$.

To complete the proof of the theorem, we apply the Mean Value Theorem to get inequalities for the slope of $f^n$ on appropriate subintervals of $[y, z]$. Four possible cases are treated separately.

*Case (i): $f^n$ increasing on $I$ with graph in $I \times (w_{i+1}, w_i)$.*

The Mean Value Theorem gives an $x_0 \in (y, a)$ such that

$$(f^n)'(x_0) = \frac{f^n(a) - f^n(y)}{a - y} = \frac{f^n(a)}{a - y} \geq \frac{w_{i+1}}{a - y} \tag{1}$$

an $x_1 \in (a, b)$ such that

$$\begin{aligned}(f^n)'(x_1) = \frac{f^n(b) - f^n(a)}{b - a} &\leq \frac{w_i - w_{i+1}}{b - a} \\ &< \frac{kw_{i+1} - w_{i+1}}{b - a} = \frac{k - 1}{b - a} w_{i+1}\end{aligned} \tag{2}$$

and finally, using the fact that $f^n(b) \leq \frac{1}{2}$, an $x_2 \in (b, m)$ such that

$$(f^n)'(x_2) = \frac{f^n(m) - f^n(b)}{m - b} = \frac{1 - f^n(b)}{m - b} \geq \frac{\frac{1}{2}}{m - b} \geq \frac{\frac{1}{2}}{z - b}. \tag{3}$$

Combining the inequalities of Step 5 with (1), (2) and (3) gives

$$(f^n)'(x_0) \geq \frac{w_{i+1}}{a - y} > \frac{k - 1}{b - a} w_{i+1} > (f^n)'(x_1)$$

and

$$(f^n)'(x_2) \geq \frac{\frac{1}{2}}{z-b} > \frac{k-1}{b-a} w_{i+1} > (f^n)'(x_1).$$

These inequalities show that the derivative of $f^n$ must have a positive local minimum at some point of $(x_0, x_2)$, so $f^n$ has a woggle.

*Case (ii): $f^n$ increasing on $I$ with graph in $I \times (1-w_i, 1-w_{i+1})$.*

Similar reasoning gives $x_0 \in (y, a), x_1 \in (a, b)$ and $x_2 \in (b, m)$ such that

$$(f^n)'(x_0) \geq \frac{\frac{1}{2}}{a-y} > \frac{k-1}{b-a} w_{i+1} > (f^n)'(x_1)$$

$$(f^n)'(x_2) \geq \frac{w_{i+1}}{z-b} > \frac{k-1}{b-a} w_{i+1} > (f^n)'(x_1)$$

giving $(f^n)'$ a positive local minimum in $(x_0, x_2)$.

*Case (iii): $f^n$ decreasing on $I$ with graph in $I \times (w_{i+1}, w_i)$.*

Here one obtains $x_0 \in (m, a), x_1 \in (a, b)$ and $x_2 \in (b, z)$ such that

$$(f^n)'(x_0) \leq \frac{-\frac{1}{2}}{a-y} < \frac{1-k}{b-a} w_{i+1} < (f^n)'(x_1)$$

$$(f^n)'(x_2) \leq \frac{-w_{i+1}}{z-b} < \frac{1-k}{b-a} w_{i+1} < (f^n)'(x_1)$$

giving $(f^n)'$ a negative local maximum in $(x_0, x_2)$.

*Case (iv): $f^n$ decreasing on $I$ with graph in $I \times (1-w_i, 1-w_{i+1})$.*

Here one obtains $x_0 \in (m, a), x_1 \in (a, b)$ and $x_2 \in (b, z)$ such that

$$(f^n)'(x_0) \leq \frac{-w_{i+1}}{a-y} < \frac{1-k}{b-a} w_{i+1} < (f^n)'(x_1)$$

$$(f^n)'(x_2) \leq \frac{-\frac{1}{2}}{z-b} < \frac{1-k}{b-a} w_{i+1} < (f^n)'(x_1)$$

giving $(f^n)'$ a negative local maximum in $(x_0, x_2)$.

Hence in every case $f^n$ has a woggle. The remaining details of the argument for cases (ii), (iii) and (iv) are left as an exercise. ■

Although we have proven the result only for symmetric maps, it is possible to modify the proof so that it works in the non-symmetric case, hence leading to the more general version of Theorem 10.3.2. Instead of using the sequence $1 - w_i$ to form our pigeon-holes in cases (ii) and (iv), we need to use the geometry of a one-hump map to find another sequence of zeroes converging to 1. This requires a fair bit of reworking at steps 4 and 5. Good luck if you try it!

## Exercises 10.4

10.4.1. Prove that the two inequalities given in equation (1) in the proof of the Theorem, yield the four given in equation (2) for each $w_{i+1}$.

[Hint: $w_{i+1} \leq \frac{1}{2}$ for all $i$.]

10.4.2. Fill in the missing details of the arguments for cases (ii), (iii) and (iv) in the proof of the Theorem.

[Hint: In cases (iii) and (iv) you will need to multiply the inequalities in (2) by $-1$ and do a little extra work with the inequality from part (d) of the proof.]

10.4.3. Adapt the technique of part (d) of the proof of the Theorem to show that $f(x) < k(1 - x)$ for all $x \in [0, 1]$ and hence

$$w_i = f(1 - w_{i+1}) < k(1 - w_{i+1})$$

for all $i$.

10.4.4. (Exercise 10.2.10 Revisited.) Suppose that $f : [0, 1] \to [0, 1]$ is a one-hump mapping with negative Schwarzian derivative and $f'(0) > 1$ and critical point $c$. Use the Mean Value Theorem to show that $f$ has no fixed points in $(0, c)$.

[Hint: Suppose there is a fixed point $p \in (0, c)$ and apply the Mean Value Theorem on the intervals $[0, p]$ and $[p, c]$. ]

## Additional reading for Chapter 10

The Schwarzian derivative was originally introduced in the study of a class of functions in complex analysis in the Ninteenth Century by Hermann Schwarz [Sch]. Most references on the use of Schwarzian derivatives in dynamical systems are rather difficult and make little effort to motivate the definition from a dynamical point of view. The Schwarzian derivative is commonly used in studying the period doubling sequence of bifurcations discussed in Section 6.4 or in connection with Singer's Theorem [Sin]. Singer's Theorem gives restrictions on the number of attracting orbits of any given period for an interval map and is also discussed in Section II.4 of [CE], Section 1.11 of [De1], Section 1.8 of [Gul] and Section II.6 of [dMvS].

Section 1.11 of [De1] gives a detailed discussion of the use of the Schwarzian derivative to prove that maps are chaotic. The fact that a negative Schwarzian derivative rules out the possibility of a positive local minimum or negative local maximum is discussed in Section 1.8 of [Gul] and Section II.6 of [dMvS].

The version of negative Schwarzian derivative we use is that discussed in [Mi].

## References

[De1] Robert L. Devaney, *An Introduction to Chaotic Dynamical Systems: Second Edition,* Addison-Wesley, Menlo Park California, 1989.

[CE] P. Collet, J.P. Eckman, *Iterated Maps of the Interval as Dynamical Systems,* Birkhäuser, Basel 1980.

[dMvS] Welington de Melo and Sebastian van Strien, *One-Dimensional Dynamics,* Penguin 1991.

[Gul] Denny Gulick, *Encounters with Chaos,* McGraw-Hill, Inc., 1992.

[Mi] John Milnor, "On the Concept of Attractor", *Communications in Mathematical Physics,* **99** (1985), 177-195.

[Sch] Hermann Schwarz, "Bestimmung einer speceiellen Minimalfläche", in: *Mathematische Abhanlungen,* Chelsea Publishing Company, New York, 1972, 6-91 (Originally published 1867).

[Sin] David Singer, "Stable Orbits and Bifurcation of Maps of the Interval", *SIAM Journal of Applied Mathematics,* **35** (1978), 260-267.

# 11

# CHANGING COORDINATES

*In this chapter we discuss some elementary ideas associated with coordinate systems on a line. Many different coordinate systems are possible. A change from one coordinate system to another is sometimes necessary. Such a change determines a mapping $h : \mathbb{R} \to \mathbb{R}$ called the transition mapping for the change.*

*A mapping $F$ of the line into itself can be considered independently of any particular coordinate system. It maps*

*a point $P$ to a point $F(P)$.*

*Given a coordinate system, however, we can introduce another mapping $f : \mathbb{R} \to \mathbb{R}$ which maps*

*the coordinate of $P$ to the coordinate of $F(P)$.*

*We call $f$ the representative of $F$ in the given coordinate system.*

*We will show that if $f$ and $g$ are the representatives of $F$ in two different coordinate systems then*

$$g = h \circ f \circ h^{-1},$$

*where $h$ is the transition mapping for the change of coordinate system. This relationship between the mappings $f$ and $g$ is called conjugacy and it will be discussed in greater generality in Chapter 12.*

## 11.1 CHANGE OF VARIABLE

As a prelude to changing coordinates, we first change the variable in a difference equation

$$x_{n+1} = f(x_n). \tag{1}$$

This difference equation may have arisen as a discrete model of population growth. In simple models $x_n$ might be the total number of *individuals* in the population, after $n$ breeding seasons.

More often, however, $x_n$ denotes *the number of individuals per unit area*; that is, *the population density.* If we change the unit of area, the size $y_n$ of the population in the new units is given by the formula

$$y_n = cx_n \tag{2}$$

where $c > 0$ is a constant whose value depends on the particular units of area involved. The formula (2) is an example of a *change of variable*: from the variable '$x_n$' to the variable '$y_n$'.

The change of variables (2) is said to be *linear* inasmuch as the graph of $y_n$ against $x_n$ is a line through the origin. Linear changes of variable occur whenever the population variable is rescaled. The aim of rescaling may be to change the units as in the above discussion; or it may be to reduce the number of parameters occuring on the right-hand side of a difference equation (as in Section 1.3).

We shall now investigate how the difference equation changes when we make such a change of variable. Along with (2) we shall need to have a formula which expresses $x_n$ in terms of $y_n$. This is obtained by solving (2) for $x_n$ to get

$$x_n = c^{-1}y_n. \tag{3}$$

Using (3) to substitute both $x_n$ and $x_{n+1}$ in (1) gives finally

$$y_{n+1} = cf(c^{-1}y_n). \tag{4}$$

To prepare for later developments, it is useful to write the change of variable (2) and its inverse (3) in terms of mappings.

A mapping $h : X \to Y$, where $X$ and $Y$ are sets, is said to be *invertible* if for each $y \in Y$ the equation

$$y = h(x)$$

has exactly one solution $x \in X$. The inverse mapping for $h$ is then defined as the mapping $h^{-1}$ with

$$h^{-1}(y) = x.$$

Let $h$ be the mapping with

$$h(x) = cx$$

so that (2) can be written as

$$y_n = h(x_n). \tag{5}$$

Solving this for $x_n$ in terms of $y_n$ gives

$$x_n = h^{-1}(y_n) \tag{6}$$

where $h^{-1}$ is the *inverse* mapping to $h$.

The difference equation for $y_n$ can now be derived as follows:

$$\begin{aligned} y_{n+1} &= h(x_{n+1}) && \text{by (5)} \\ &= h(f(x_n)) && \text{by (1)} \\ &= h(f(h^{-1}(y_n))) && \text{by (6)} \\ &= (h \circ f \circ h^{-1})(y_n) \end{aligned}$$

Thus the difference equation for $y_n$ is

$$y_{n+1} = g(y_n) \tag{7}$$

where

$$g = h \circ f \circ h^{-1}. \tag{8}$$

Thus we have shown: *to use the new variables $y_n = h(x_n)$, replace right-hand side $f$ of the original difference equation by the right-hand side $g$.*

Since merely changing the unit of area can have no effect on the growth of the actual population, we expect the overall pattern of solutions of the two difference equations to be the same.

Periodic solutions of one equation will match with periodic solutions of the other, asymptotic solutions will match with asymptotic solutions, and so on. We express this by saying that the two difference equations (1) and (7) must have *the same dynamics*.

## Exercises 11.1

11.1.1. Let $x_n$ be the number of individuals per *acre* and $y_n$ the number per *hectare*. Given that there are 2.471 acres to a hectare, find the constant $c$ such that $y_n = cx_n$.

11.1.2. Let $h : \mathbb{R} \to \mathbb{R}$ where $h(x) = cx$ for some $c > 0$.

(a) Find the inverse of $h$.

(b) Hence find a formula for $g(y)$ where $g = h \circ f \circ h^{-1}$.

(c) Hence check that the difference equation (7) in the text is the same as (4).

11.1.3. Show that under the change of variable $y_n = x_n^2$ (where $x_n \geq 0$) the difference equation

$$x_{n+1} = \sqrt{\mu} x_n \sqrt{1 - x_n^2}$$

becomes the logistic equation $y_{n+1} = \mu y_n(1 - y_n)$.

11.1.4. Let $h : [0, \infty) \to [0, \infty)$ with $h(x) = x^2$. Show that $h$ is invertible and find the inverse mapping $h^{-1}$. Discuss the relevance of this to the change of variables used in Exercise 3.

## 11.2 COORDINATES

We now discuss the idea of a coordinate system on a line and we interpret a change of variable $y = h(x)$ as a change of coordinate on the line.

A coordinate system allows us to match each point $P$ on a line with a real number $x$. Every point on the line gets matched with a different real number, and no real numbers are left out. Thus a coordinate system sets up a one-to-one correspondence $P \leftrightarrow x$ between a set of points and a set of numbers.

A *Cartesian coordinate system* can be specified by giving two distinct points on the line, say $P_0$ and $P_1$. The point $P_0$ is taken as *origin* and the *unit of length* is taken as the length of the segment between $P_0$ and $P_1$.

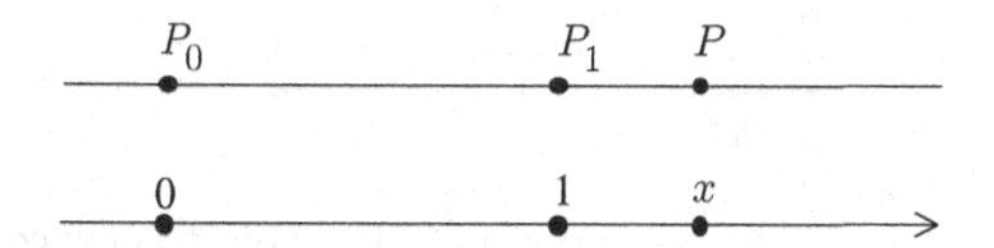


**Figure 11.2.1.** A Cartesian coordinate system.

The coordinate of a point $P$ on the line is then defined as the number

$$x = \pm \; distance \; from \; P \; to \; P_0$$

The plus sign is chosen if and only if $P$ is on the same side of the origin as $P_1$. Hence the coordinates of $P_0$ and $P_1$ are 0 and 1 respectively.

Two coordinate systems for the line are said to have the same *orientation* if in both coordinate systems, the point 1 is on the same side of 0.

### Changing coordinates

Let a typical point $P$ on a line have coordinate $x$ relative to one coordinate system and $y$ relative to a second coordinate system. Each $x$ determines a unique $P$, which in turn determines a unique $y$. Hence

$$y = h(x)$$

for some function $h$ that is uniquely determined by the two coordinate systems. Similarly $x$ is determined uniquely by $y$ so we can write

$$x = h^{-1}(y).$$

Thus each change of coordinates corresponds to a map $h$ with inverse $h^{-1}$. The mapping $h$ is called the *transition mapping* from the first coordinate system to the second.

**11.2.1 Example** *Two cartesian coordinate systems for a line are shown in Figure 11.2.2. They have the same origin, the same orientation and the unit of length in the new coordinate system is half that in the old one. Find the transition mapping $h$ and its inverse $h^{-1}$.*

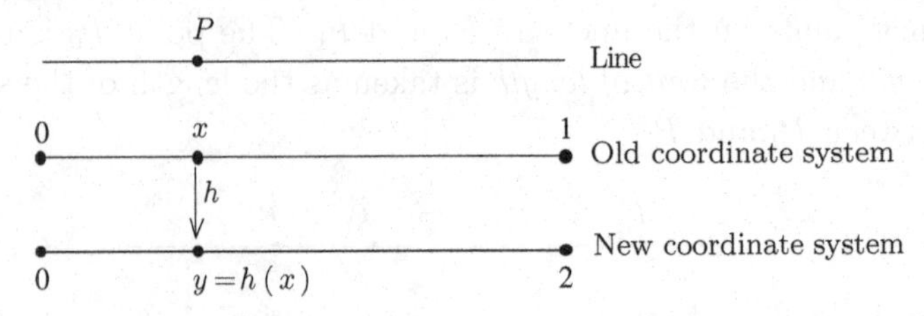


**Figure 11.2.2.** The transition map from old to new coordinates.

*Solution:* If $x$ and $y$ are the coordinates of the same point $P$ referred to the old and the new coordinate systems respectively, then $y = 2x$ and so

$$h(x) = 2x.$$

Solving for $x$ in terms of $y$ gives $x = \frac{1}{2}y$. Hence $h^{-1}(y) = \frac{1}{2}y$. ■

**11.2.2 Example** *Two cartesian coordinate systems for the line are shown in Figure 11.2.3. They have different orientation and the new unit of length is half the old one. The new origin is one (old) unit to the right of the old origin. Find the transition mapping $h$ and give its inverse $h^{-1}$.*

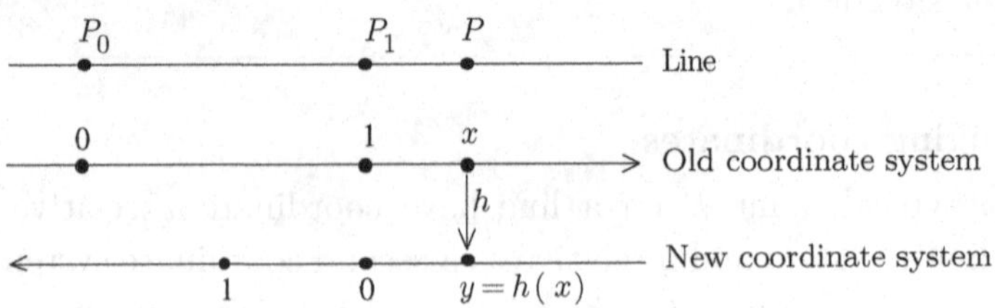

**Figure 11.2.3** A change of unit, orientation and origin.

*Solution:* Let $x$ and $y$ be the coordinates of the same point $P$ referred to the old and new coordinate systems respectively. If $x > 1$ then

$$\begin{aligned} \textit{distance from } P_1 \textit{ to } P &= (x-1) \quad \textit{old units} \\ &= 2(x-1) \quad \textit{new units} \end{aligned}$$

But this distance is also equal to $-y$ *new units.* Hence

$$y = -2x + 2.$$

A similar argument gives the same result in the cases where $x \leq 1$. Thus the transition mapping $h$ is given by

$$h(x) = -2x + 2.$$

The inverse mapping is given by $h^{-1}(y) = -\frac{1}{2}y + 1$. ■

## Exercises 11.2

11.2.1. Figure 11.2.4 shows details for coordinate system 1 for the line. Coordinate system 2 has the following property: *it has the same origin and the same unit of length as the first coordinate system, but it has the opposite orientation.*

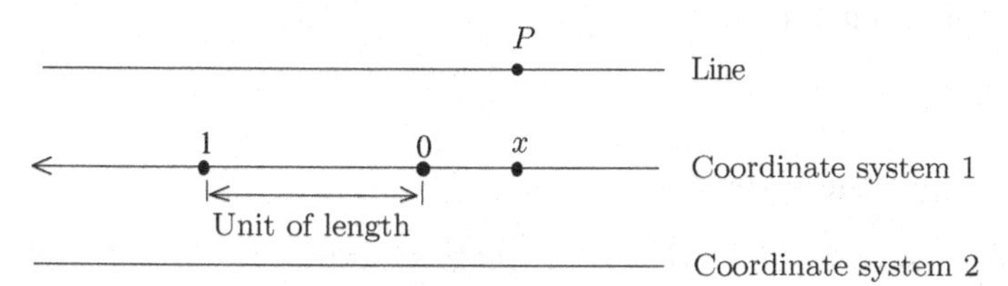


**Figure 11.2.4.**

(a) Show the points 0 and 1 for the second coordinate system.

(b) Let $y$ be the coordinate of the point $P$ in the second coordinate system. Show $y$ on the last line in your sketch and then express $y$ in terms of $x$.

(c) Give the transition mapping $h : \mathbb{R} \to \mathbb{R}$ for the change of coordinates.

11.2.2. Repeat Exercise 1, but now let the second coordinate system have the following property:

*It has the same unit of length and the same orientation as the first coordinate system, but the origin is shifted one unit to the right of that in the first coordinate system.*

11.2.3. Repeat Exercise 1, but now let the second coordinate system have the following property:

*It has the same origin and the same orientation as the first coordinate system, but the unit of length is twice that in the first coordinate system.*

11.2.4. Figure 11.2.5 shows details for the first coordinate system for the line.

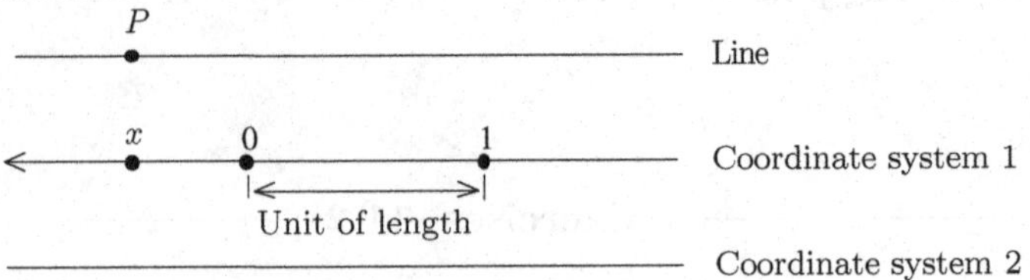


**Figure 11.2.5.**

The transition mapping $h$ from the first coordinate system to the second is given by

$$h(x) = \frac{1}{2}x - \frac{1}{2}.$$

(a) Find the inverse of $h$.

(b) Describe the second coordinate system.

## 11.3 MAPPINGS ON THE LINE

Some mappings from a line to itself are so simple that they can be described without reference to any coordinate system. More complicated mappings, however, are most easily specified with the aid of a coordinate system. The following definition, which is illustrated in Figure 11.3.1, provides the relevant terminology.

**11.3.1 Definition** Let $F$ map a line into itself so that it maps a point $P$ to a point $F(P)$. In a given a coordinate system for the line we can define a mapping $f : \mathbb{R} \to \mathbb{R}$ as the *representative of* $F$ so that f maps the coordinate of $P$ to the coordinate of $F(P)$. ∎

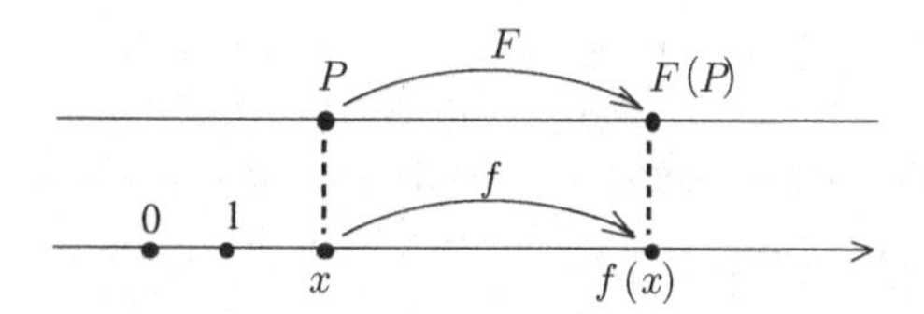


**Figure 11.3.1** $F$ maps points. Its representative $f$ maps their coordinates.

**11.3.2 Example** *Let $A$ be a point on the line. Let $F$ be the mapping which assigns to each point $P$ its reflection in the point $A$ (as in Figure 11.3.2). Find the representative of $F$ in a cartesian coordinate system for the line.*

*Solution:* Suppose that $A$ has coordinate $a \in \mathbb{R}$. Let $f : \mathbb{R} \to \mathbb{R}$ denote the representative of $F$ in the coordinate system. From Figure 11.3.2 it follows that $f(x) - a = a - x$, and so $f(x) = 2a - x$. ∎

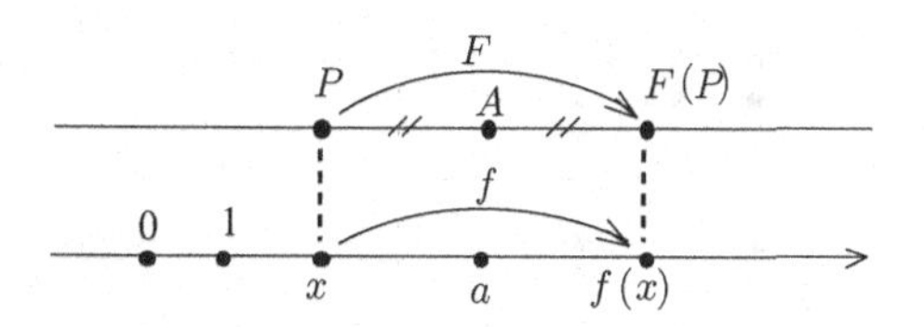


**Figure 11.3.2** $F$ is a reflection in the point $A$.

A neater solution to the above example is to choose the point $A$ as the origin of the coordinate system.

This choice gives immediately $f(x) = -x$ as shown in Figure 11.3.3.

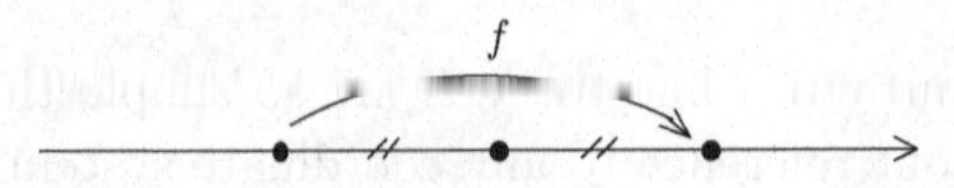


**Figure 11.3.3** How to choose the origin.

The above example thus illustrates an important principle: *a sensible choice of coordinate system which fits in with the underlying geometry of the problem can often simplify subsequent calculations.*

## Changing the mappings

The representative of a given mapping $F$ of a line, with respect to a coordinate system, will change with the coordinate system. The following example shows how the transition mapping can be used to express the new representative of $F$ in terms of the old one. We assume that $F$ has only a line segment as its domain and codomain. Hence its representatives have intervals for their domains and codomains.

**11.3.3 Example** *Let $F$ map a line segment into itself. Suppose that, with respect to two different cartesian coordinate systems, the representatives of $F$ are $f : [a, b] \to [a, b]$ and $g : [c, d] \to [c, d]$ respectively.*

*Show that $h \circ f = g \circ h$ where $h : [a, b] \to [c, d]$ is the transition mapping for the change of coordinates.*

*Solution:* Figure 11.3.4 shows the relationship between the mappings.

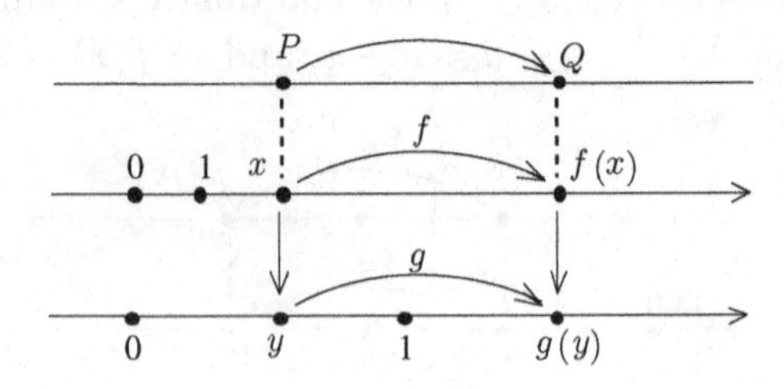


**Figure 11.3.4.**

First, since $h$ changes coordinates in system 1 into coordinates in system 2, we see from Figure 11.3.4 that

$$g \text{ maps } h(x) \text{ to } h(f(x))$$

Second, since $y = h(x)$, we see that

$$g \text{ maps } h(x) \text{ to } g(h(x)).$$

Equating the two rightmost expressions gives $h(f(x)) = g(h(x))$ so $h \circ f = g \circ h$ by the definition of the equality of two functions. ■

## Changing the graphs

The relationship $h \circ f = g \circ h$ can also be motivated through the graphs of the mappings $f$ and $g$.

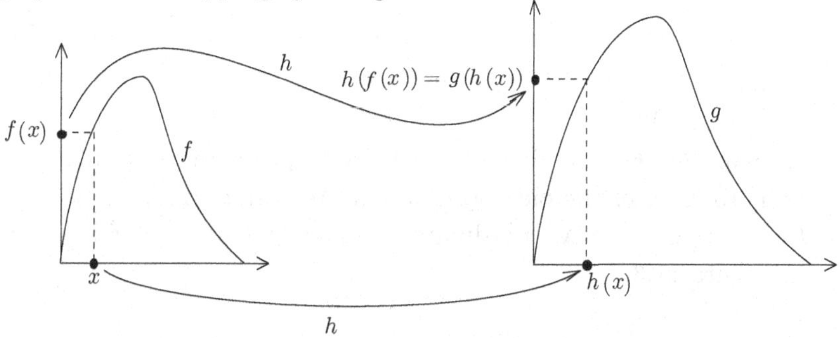


**Figure 11.3.5.**

In Figure 11.3.5, the original mapping of the line, when the first coordinate system is used,

$$\text{maps } x \text{ to } f(x).$$

But the mapping $h$ when applied to these numbers, gives the new coordinates. In the new coordinates, the original mapping

$$\text{maps } h(x) \text{ to } h(f(x)).$$

But in the new coordinates the original mapping becomes $g$, so it also

$$\text{maps } h(x) \text{ to } g(h(x)).$$

Hence, as before,

$$h(f(x)) = g(h(x)). \tag{2}$$

The equality (2) can also be written (see Exercise 11.3.4) in the form

$$g = h \circ f \circ h^{-1}. \tag{3}$$

## Exercises 11.3

11.3.1. Let $h : [0,1] \to [-1,0]$ with $h(x) = -x$ be the transition map from an old coordinate system to a new one. The old coordinate system is shown in the following figure.

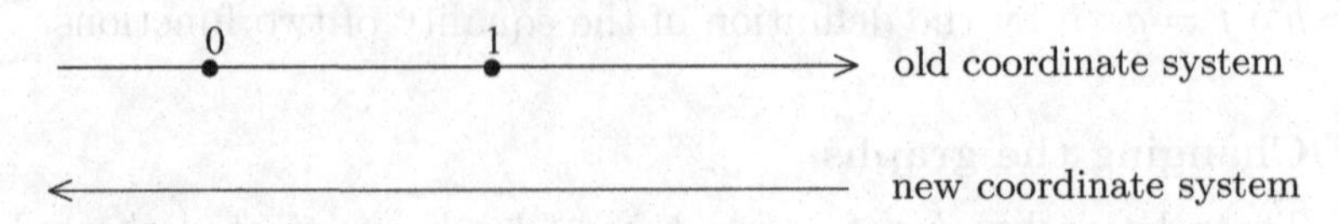


**Figure 11.3.6**

(a) Copy the figure onto your page and then show, directly beneath it, the new coordinate system (origin, unit of length, orientation).

A mapping $F$ of a line segment has representative $f$ mapping $[0,1]$ to $[0,1]$ in the old coordinate system and $g$ mapping $[-1,0]$ to $[-1,0]$ in the new coordinate system. The graph of $f$ is shown in Figure 11.3.7.

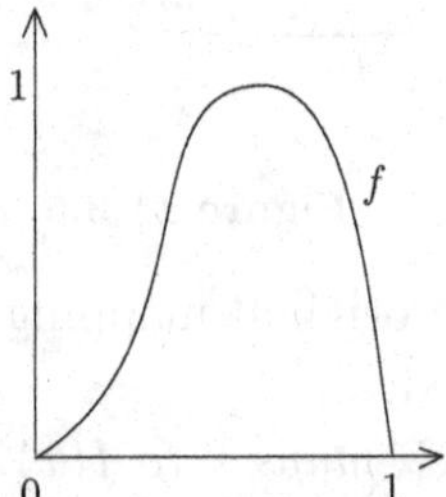


**Figure 11.3.7**

(b) Sketch the graph $g$ using the procedure suggested in Figure 11.3.5 in the text.

(c) Find the inverse of the mapping $h$ and then use the relationship $g = h \circ f \circ h^{-1}$to find an explicit formula for $g(x)$.

11.3.2. Repeat Exercise 1, but now let $h(x) = x - 1$.

11.3.3. Repeat Exercise 1, but now let $h : [0,1] \to [0, \frac{1}{2}]$ with $h(x) = \frac{1}{2}x$ and with $g$ mapping $[0, \frac{1}{2}]$ to $[0, \frac{1}{2}]$.

11.3.4. Let $f$ and $g$ be the functions whose graphs appear below.

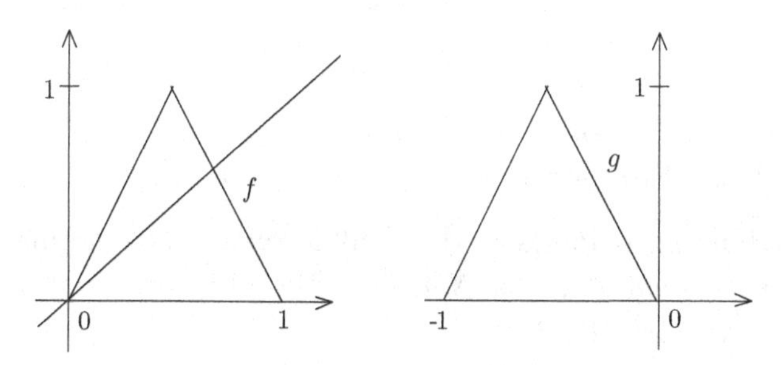


**Figure 11.3.8**

(a) Do the mappings $f$ and $g$ have similar dynamics?

(b) Is there a mapping $h$, describing a change of coordinates, such that $g = h \circ f \circ h^{-1}$ ?

11.3.5. Let $S$ and $T$ be sets and let $f : S \to S$ and let $g : T \to T$. Let $h : S \to T$ be invertible.

(a) Show that $h \circ h^{-1}$ is the identity map id on the set $T$.

(b) Verify that the relation $h \circ f = g \circ h$ is equivalent to $g = h \circ f \circ h^{-1}$.

11.3.6. Let $F$ be a mapping from a line into itself such that for each point $P$ on the line $F(P) = P$. Prove that if $f : \mathbb{R} \to \mathbb{R}$ is the representative of $F$ in a coordinate system for the line, then $f(x) = x$ for all $x \in \mathbb{R}$.

---

## Additional reading for Chapter 11

In Section 5.11 of [PMJPSY], the concept of changing coordinates is used so that the graph of the tent map can be transformed into the graph of the logistic map and to show similar dynamics between them. But the idea of changing coordinates in general, is not explained.

Section 10.6 of [PJS] gives the same discussion about the change of coordinates as in [PMJPSY].

Section 3.5 of [Ba] describes in detail a general change of coordinates for a given space. However, the difference between points and coordinates is not very clear.

---

## References

[Ba] Michael F. Barnsley, *Fractals Everywhere,* 2nd edition, Academic Press Professional, Boston 1993.

[PJS] Hans-Otto Peitgen, Hartmut Jürgens and Dietmar Saupe, *Chaos and Fractals, New Frontiers of Science*, Springer-Verlag, New York, 1992.

[PMJPSY] Hans-Otto Peitgen, Evan Maletsky, Hartmut Jürgens, Terry Perciante, Dietmar Saupe, and Lee Yunker, *Fractals for the Classroom: Strategic Activities Volume Two,* Springer-Verlag, 1992. (two volumes)

# 12

# CONJUGACY

*In the previous chapter we described some simple changes of coordinates on the line in terms of invertible affine mappings. These mappings are called transition mappings from one coordinate system to the other.*

*In this chapter we consider changes of coordinates in which the transition mappings are not necessarily affine, but are still invertible and continuous.*

*Topological conjugacy (or just conjugacy for short) is the relationship which exists between two mappings when one is obtained from the other by a change of coordinates. We shall prove that when two mappings are in this relationship, they have similar dynamics; in particular when one is chaotic so is the other.*

*Unfortunately, there is no systematic method for showing conjugacy between a given pair of mappings and it is usually difficult to prove it. We can prove, however, that the tent mapping $T_4$ and the logistic mapping $Q_4$ are conjugate. More generally, we prove that all maps with wiggly iterates are conjugate to the tent map.*

## 12.1 DEFINITION AND EXAMPLES

This section is about changes of coordinates between two mappings which are no longer assumed to be affine. A transition mapping for any change of coordinates needs to be both continuous and invertible. Continuity ensures that points which are close together in one coordinate system will also be close in the other coordinate system. Invertibility ensures that we can 'undo' the change of coordinate.

**12.1.1 Definition** Let $I$ and $J$ be two intervals. A continuous mapping $h : I \to J$ with continuous inverse $h^{-1} : J \to I$ is called a *homeomorphism.* ■

Thus the sorts of mappings which we wish to consider as possible transition mappings for a change of coordinate system are homeomorphisms.

**12.1.2 Definition** A mapping $f : I \to I$ is *topologically conjugate* to a mapping $g : J \to J$ if there is a homeomorphism $h : I \to J$ such that

$$h \circ f = g \circ h.$$

■

Figure 12.1.1 shows a convenient way to visualize the conjugacy relation. The diagram says that *applying f then h* is the same as *applying h then g* and for this reason the diagram is called *commutative.* It makes clear the relationship of the mappings to their domains and codomains.

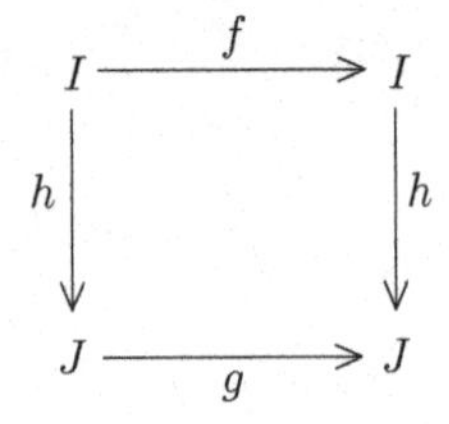


**Figure 12.1.1.** Commutative diagram showing conjugacy of $f$ to $g$.

We indicate the role of the homeomorphism $h$ by saying that $f$ is conjugate to $g$ via $h$, or that $h$ is a *topological conjugacy* from $f$ to $g$, or that $h$ is a conjugacy between $f$ and $g$. *Topological conjugacy* is often abbreviated to just *conjugacy*.

**12.1.3 Example** *Show that if $f$ is conjugate to $g$ via $h$, then $f^2$ is conjugate to $g^2$ via $h$.*

*Solution:* We are to show that if $h \circ f = g \circ h$, then $h \circ f^2 = g^2 \circ h$. Let $h \circ f = g \circ h$ and then compose on the right with $f$ to get

$$h \circ f^2 = g \circ h \circ f = g \circ g \circ h = g^2 \circ h. \qquad \blacksquare$$

**12.1.4 Lemma** *Let $n \in \mathbb{N}$. If $f$ is conjugate to $g$ via $h$, then $f^n$ is conjugate to $g^n$ via $h$.*

*Proof:* This is set as Exercise 12.1.10. ■

To prove that $f$ is conjugate to $g$, we must find a homeomorphism $h$ which satisfies the conjugacy relation. There are no rules for finding such a homeomorphism. Instead we must try examples until we get one that works. We can draw on the many examples of homeomorphisms which are included as standard fare in calculus courses.

> The function $h : \mathbb{R} \to \mathbb{R}^+$ with $h(x) = e^x$ is a homeomorphism. Its inverse is $h^{-1} : \mathbb{R}^+ \to \mathbb{R}$ with $h^{-1}(x) = \log_e(x)$.
>
> The function $h : \mathbb{R} \to \mathbb{R}$ with $h(x) = x^{\frac{1}{3}}$ is a homeomorphism. Its inverse is $h^{-1} : \mathbb{R} \to \mathbb{R}$ with $h^{-1}(x) = x^3$.
>
> The function $h : [0, \infty) \to [0, \infty)$ with $h(x) = x^{\frac{1}{2}}$ is a homeomorphism. Its inverse is $h^{-1} : [0, \infty) \to [0, \infty)$ with $h^{-1}(x) = x^2$.

We shall now use these examples of homeomorphisms to prove conjugacy between various pairs of mappings.

**12.1.5 Example** *Show that $f : \mathbb{R} \to \mathbb{R}$ with $f(x) = -x$ is conjugate to $g : \mathbb{R}^+ \to \mathbb{R}^+$ with $g(x) = 1/x$.*

*Solution:*

We want a homeomorphism $h : \mathbb{R} \to \mathbb{R}^+$ such that $h(f(x)) = g(h(x))$; that is

$$h(-x) = \frac{1}{h(x)}.$$

Thus we want $h$ to convert negatives into reciprocals.

This suggests the exponential function.

Choose $h : \mathbb{R} \to \mathbb{R}^+$ with $h(x) = e^x$.
Hence $h$ is a homeomorphism. It is, furthermore, a conjugacy from $f$ to $g$ since

$$h(f(x)) = e^{-x} = \frac{1}{e^x} = g(h(x)). \qquad \blacksquare$$

**12.1.6 Example** *Show that $f : [0, \infty) \to [0, \infty)$ with $f(x) = \frac{1}{4}x$ is conjugate to $g : [0, \infty) \to [0, \infty)$ with $g(x) = \frac{1}{2}x$.*

*Solution:*

We want a homeomorphism $h : [0, \infty) \to [0, \infty)$ such that $h(f(x)) = g(h(x))$; that is

$$h\left(\frac{1}{4}x\right) = \frac{1}{2}h(x).$$

Thus we want $h$ to convert $\frac{1}{4}$ into its square root.

This suggests the square root function.

Choose $h : [0, \infty) \to [0, \infty)$ with $h(x) = \sqrt{x}$. Hence $h$ is a homeomorphism. It is a conjugacy from $f$ to $g$ since for all $x > 0$,

$$h(f(x)) = \sqrt{\frac{1}{4}x} = \frac{1}{2}\sqrt{x} = g(h(x)). \qquad \blacksquare$$

**Fig. 12.1.2** Graph of $h(x) = \sqrt{x}$.

This is the graph of the homeomorphism $h$ used above in Example 12.1.6. It is not differentiable at the origin because of the vertical tangent.

Thus Example 12.1.6 explains why we do not impose the condition of differentiability on the homeomorphisms used in conjugacy.

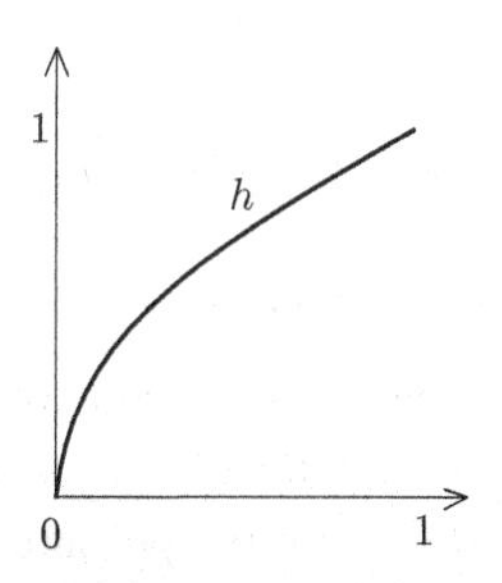


Exercise 12.1.5 shows how to construct a conjugacy between $f$ and $g$ when these mappings are given by the same formulae as in the above example, but have the whole of $\mathbb{R}$ for domain and codomain. The following example shows, nevertheless, that there is no conjugacy $h$ between $f$ and $g$ such that both $h$ and $h^{-1}$ are *differentiable.*

**12.1.7 Example** *Let $f$ and $g$ map $\mathbb{R}$ into itself where $f(x) = \frac{1}{4}x$ and $g(x) = \frac{1}{2}x$. If $f$ is conjugate to $g$ via a homeomorphism $h$, then either $h$ or $h^{-1}$ is not differentiable at* 0.

*Solution:* If $h$ is differentiable then $h\left(\frac{1}{4}x\right) = \frac{1}{2}h(x)$ and so

$$\frac{1}{4}h'\left(\frac{1}{4}x\right) = \frac{1}{2}h'(x)$$

Hence

$$\frac{1}{4}h'(0) = \frac{1}{2}h'(0)$$

and so

$$h'(0) = 0.$$

Hence $h^{-1}$ cannot be differentiable.

Similarly we can show that if $h^{-1}$ is differentiable then $h$ cannot be differentiable. ∎

We now show that the tent mapping $T_4$ is conjuate to the logistic mapping $Q_4$ via the mapping $h : [0,1] \to [0,1]$ with $h(x) = \sin^2(x\pi/2)$. The formula for $h(x)$ is suggested by the change of variables given in Exercise 6.1.9.

**Fig. 12.1.3.** Graph of $h(x) = \sin^2(x\pi/2)$.

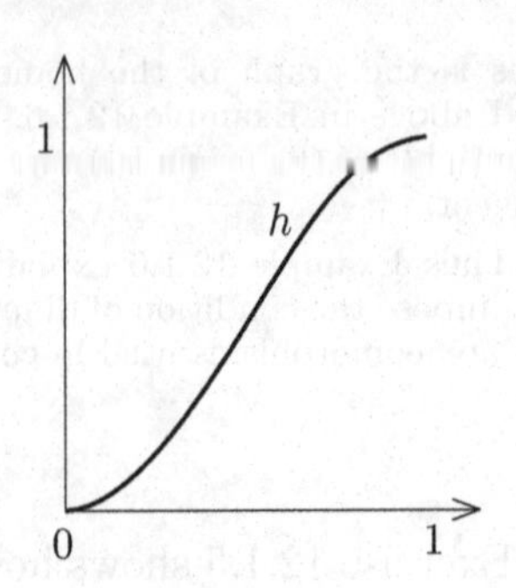


(Recall that $\sin^2$ means the *product* $\sin \cdot \sin$.)

The mapping $h$ has a derivative which is defined throughout the closed interval $[0, 1]$ and strictly positive on the open interval $(0, 1)$. It follows from standard results in calculus that $h$ is a homeomorphism.

**12.1.8 Example** *Show that the tent map $T_4$ is conjugate to the logistic map $Q_4$ via the homeomorphism $h$ given above.*

*Solution:* By definition of $T_4$,

$$h(T_4(x)) = \begin{cases} h(2x) = \sin^2(x\pi) & \text{if } 0 \le x \le \frac{1}{2} \\ h(2-2x) = \sin^2\left(\frac{(2-2x)\pi}{2}\right) = \sin^2(x\pi) & \text{if } \frac{1}{2} < x \le 1 \end{cases}$$

while

$$Q_4(h(x)) = 4h(x)(1 - h(x)) = \sin^2(x\pi)$$

Thus, in both cases

$$h(T_4(x)) = Q_4(h(x))$$ ∎

## Exercises 12.1

12.1.1. Let $f : \mathbb{R} \to \mathbb{R}$ and $g : \mathbb{R}^+ \to \mathbb{R}^+$. In each of the following cases show that $f$ is conjugate to $g$ under a suitable homeomorphism $h : \mathbb{R} \to \mathbb{R}^+$.

(a) $f(x) = 2x$ and $g(x) = x^2$.
(b) $f(x) = -2x$ and $g(x) = 1/x^2$.

12.1.2. Let $f$ and $g$ map $\mathbb{R}$ into $\mathbb{R}$ and let $f(x) = 2x$ and $g(x) = x^2$. Note that $f$ and $g$ have the same formulae as in Exercise 12.1.1(b) but $g$ has a different domain and codomain.

(a) Do the mappings $f$ and $g$ have similar dynamics?

(b) Would you expect $f$ to be conjugate to $g$? Give reasons.

12.1.3. Let $f$ and $g$ map $\mathbb{R}$ into $\mathbb{R}$ and let $f(x) = 8x$ and $g(x) = 2x$. Show that $f$ is conjugate to $g$. [Hint: Note that $2^3 = 8$.]

12.1.4. Let $f$ and $g$ map $\mathbb{R}$ into $\mathbb{R}$ and let $f(x) = 2x$ and $g(x) = \frac{1}{2}x$.

(a) Do $f$ and $g$ have similar dynamics?

(b) Is $f$ conjugate to $g$? Give reasons.

12.1.5. Let $f$ and $g$ each map $\mathbb{R}$ into $\mathbb{R}$ with $f(x) = \frac{1}{4}x$ and $g(x) = \frac{1}{2}x$.

Let $h : \mathbb{R} \to \mathbb{R}$ with $h(x) = \begin{cases} \sqrt{x} & \text{if } x \geq 0 \\ -\sqrt{-x} & \text{if } x < 0. \end{cases}$

Show that $f$ is conjugate to $g$ via $h$.

12.1.6. Let $\phi : [0, \infty) \to [1, \infty)$ be a homeomorphism.

Let $h : \mathbb{R}$ into $\mathbb{R}^+$ with $h(x) = \begin{cases} \phi(x) & \text{if } x \geq 0 \\ \frac{1}{\phi(-x)} & \text{if } x < 0. \end{cases}$

(a) Explain why $h$ is a homeomorphism.

(b) Show that $f$ and $g$ as in Example 12.1.5 are conjugate via the above $h$.

(c) How many different homeomorphisms are there under which $f$ is conjugate to $g$?

12.1.7. Let $f$ and $g$ map $\mathbb{R}$ into $\mathbb{R}$ and let $f(x) = mx$ and $g(x) = nx$ where $m$ and $n$ are real numbers. Discuss when $f$ is conjugate to $g$.

An *equivalence relation* $\sim$ on a set $F$ is a relation such that for all $a, b, c$ in $F$:

(i) $a \sim a$.

(ii) $a \sim b \Rightarrow b \sim a$.

(iii) $a \sim b$ and $b \sim c \Rightarrow a \sim c$.

12.1.8. Prove that topological conjugacy is an equivalence relation; that is, show that if $f$ *is conjugate to* $g$ is denoted by '$f \sim g$', then $\sim$ is an equivalence relation.

12.1.9. Prove that if $f$ and $g$ are conjugate via $h$ then $f^3$ and $g^3$ are conjugate via $h$.

[ Hint. Use the result of Example 12.1.3].

12.1.10. Let $n \in \mathbb{N}$. Prove that if $f$ is conjugate to $g$ via $h$, then $f^n$ is conjugate to $g^n$ via $h$.

12.1.11. Consider the following family of mappings, discussed in Section 1.3,

$$f_{\lambda,\alpha}(x) = \lambda x e^{-\alpha x}$$

where $\lambda$ and $\alpha$ are positive real parameters.

(a) Sketch the graphs of $f_{10,5}$ and $f_{10,10}$. Is it geometrically obvious that these two maps have the same dynamics? Why?

(b) Show that for any fixed $\lambda$, a pair of members $f_{\lambda,\alpha}$ and $f_{\lambda,\beta}$ are conjugate via $h(x) = \frac{\alpha}{\beta}x$.

## 12.2 APPROXIMATING A CONJUGACY

Let $f : [0,1] \to [0,1]$ and $g : [0,1] \to [0,1]$ be one-hump mappings with wiggly iterates. In this section we start from the assumption that $h$ conjugates $f$ to $g$ and concern ourselves with the practical problem of how to sketch a good approximation to the graph of the conjugacy $h$. The more theoretical problem of verifying that the method works is left until later.

### Describing the method

We call our method for approximating the conjugacy $h$ the method of *Matching Zeroes* since it involves matching zeroes of higher iterates of $f$ with those of $g$. The matching is possible because $f^n$ and $g^n$ have the same number ($\ell = 2^{n-1} + 1$) of zeroes.

**Matching Zeroes**

STEP 1. Choose $n \in \mathbb{N}$ and then plot the graphs of $f^n$ and $g^n$.

STEP 2. Mark on the graphs the zeroes of $f^n$ say

$$z_1 < z_2 < \cdots < z_\ell$$

and those of $g^n$ say

$$w_1 < w_2 < \cdots < w_\ell.$$

STEP 3. Plot the points $(z_i, w_i)$ for $1 \leq i \leq \ell$.

STEP 4. Join these points with line segments to get the graph of a piecewise affine mapping $h_n$.

The points $(z_i, w_i)$ lie on the graph of $h$, as we shall prove later.[1]

This gives a large number of points on the graph of $h$, when $n$ is large. The assumed wiggly iterates for $f$ and $g$ will ensure that the gaps between these points are small. Hence joining these points with line segments should give a good approximation to the graph of $h$.

[1] For the present, note that the zeroes of the iterates play an important part in the dynamics of the mapping. Hence the conjugacy $h$ should map the zeroes of $f^n$ to the zeroes of $g^n$ ; this means that $w_i = h(z_i)$ and hence that the point $(z_i, w_i)$ lies on the graph of $h$.

**12.2.1 Definition** We call the mapping $h_n$, in Matching Zeroes, *the nth piecewise affine approximation* to the homeomorphism $h$. ■

**12.2.2 Example** *Let $f$ be the tent mapping $T_4$ and let $g$ be the logistic mapping $Q_4$. By using the method of Matching Zeroes with $n = 1$, find some points on the graph of the conjugacy $h$ from $f$ to $g$. Hence sketch the graph of the approximation $h_1$ to $h$.*

*Solution:* In this case $n = 1$ and so $\ell = 2$. Hence we show the zeroes $z_1$ and $z_2$ of $f$ and the zeroes $w_1$ and $w_2$ of $g$ in Figure 12.2.1.

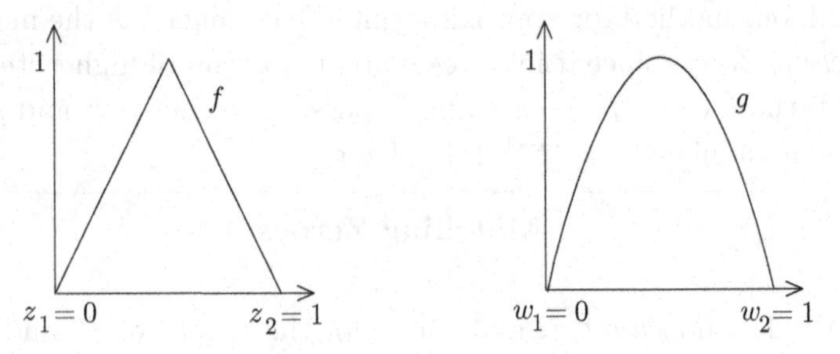


**Figure 12.2.1.** The zeroes of $f$, and also of $g$, are 0 and 1.

Next we match zeroes of $f$ with the zeroes of $g$ to get two points

$$(z_1, w_1) \text{ and } (z_2, w_2)$$

on the graph of $h$. Finally, join these points with a line segment to get the graph of $h_1$ (see Figure 12.2.2). ■

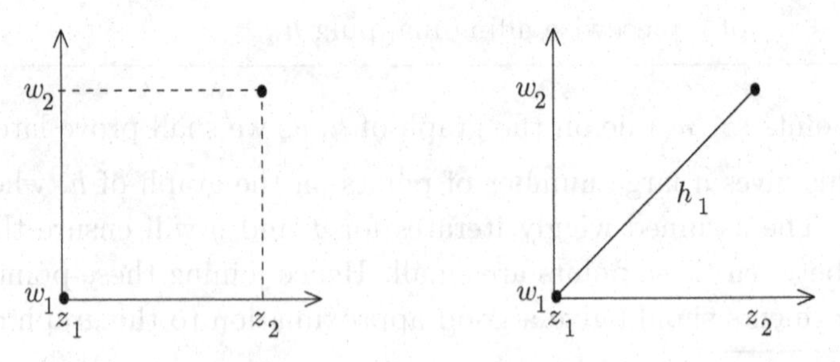


**Figure 12.2.2.** The two points on the graph of $h$ are shown on the left. The graph of $h_1$, the first piecewise affine approximation to $h$, is on the right.

**12.2.3 Example** *Let $f$ be the tent mapping $T_4$ and let $g$ be the logistic mapping $Q_4$. Use the method of Matching Zeroes with $n = 2$, to find some points on the graph of the conjugacy $h$ from $f$ to $g$. Hence sketch the graph of the approximation $h_2$ to $h$.*

*Solution:* In this case $n = 2$ and so $\ell = 3$. Hence in Figure 12.2.3 we show the zeroes $z_1$, $z_2$ and $z_3$ of $f^2$ and the zeroes $w_1$, $w_2$ and $w_3$ of $g^2$.

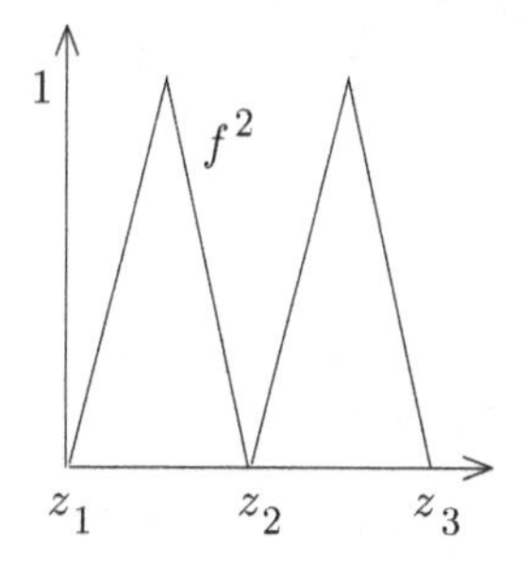

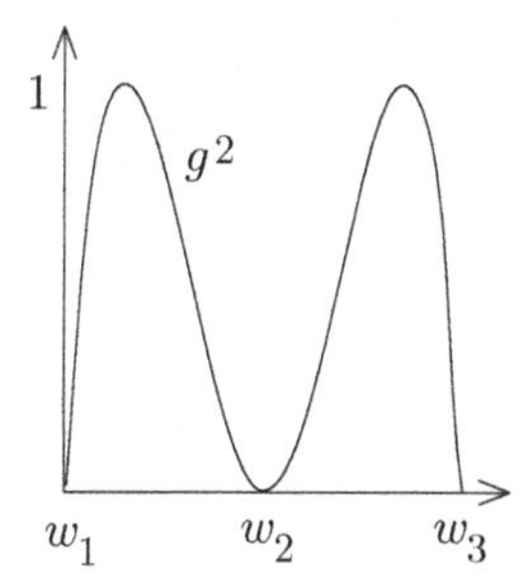


**Figure 12.2.3.** The zeroes of $f^2$, and also of $g^2$, are 0, 1/2 and 1.

Next we match the zeroes of $f^2$ with the zeroes of $g^2$ to get the three points

$$(z_1, w_1), (z_2, w_2) \text{ and } (z_3, w_3)$$

on the graph of $h$. Finally, we join pairs of consecutive points with line segments to get the graph of $h_2$ (see Figure 12.2.4). ∎

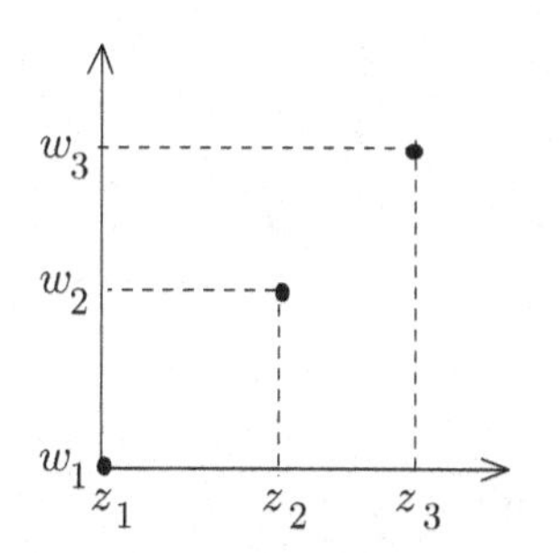

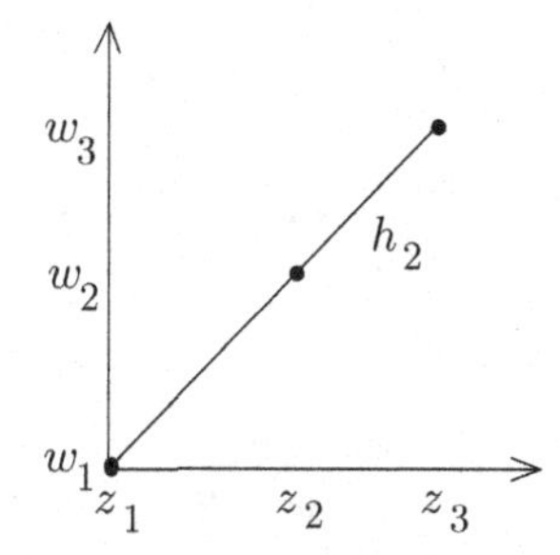


**Figure 12.2.4.** Three points on the graph of $h$ are shown on the left. The graph of $h_2$, the second piecewise affine approximation to $h$, is on the right.

**12.2.4 Example** *Let $f$ be the tent mapping $T_4$ and let $g$ be the logistic mapping $Q_4$. Use the method of Matching Zeroes with $n = 3$ to find some points on the graph of the conjugacy $h$ from $f$ to $g$. Hence sketch the graph of the approximation $h_3$ to $h$.*

*Solution:* From $n = 3$ follows $\ell = 5$. Hence in Figure 12.2.5 we show the zeroes $z_1, z_2, \ldots, z_5$ of $f^3$ and the zeroes $w_1, w_2, \ldots, w_5$ of $g^3$.

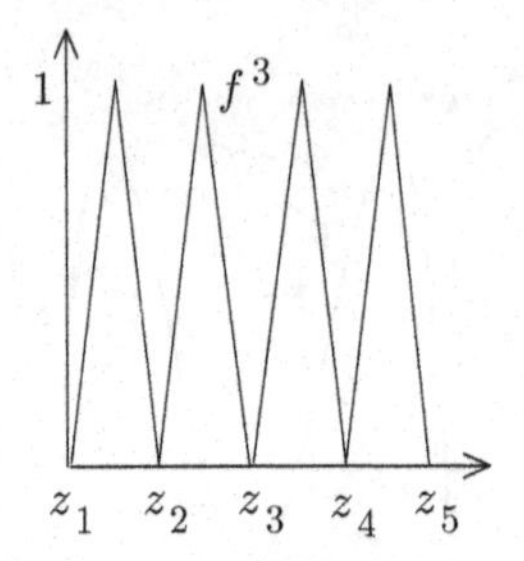


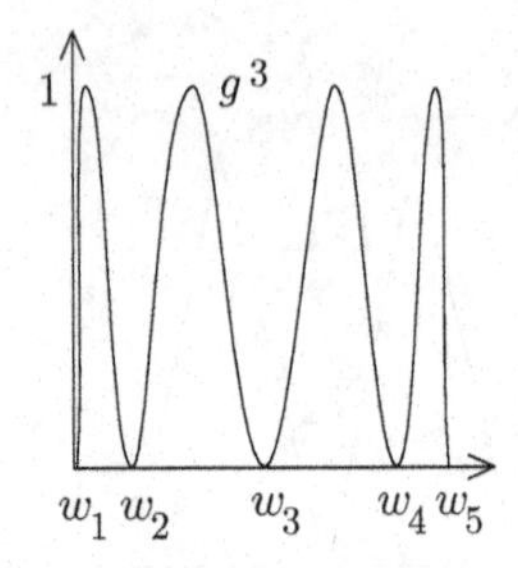


**Figure 12.2.5.** The zeroes of $f^3$ and $g^3$.

Next we match the zeroes of $f^3$ with the zeroes of $g^3$ to get five points

$$(z_1, w_1), (z_2, w_2), \ldots, (z_5, w_5)$$

on the graph of $h$. Finally, join pairs of consecutive points with line segments to get the graph of $h_3$ (see Figure 12.2.6). ∎

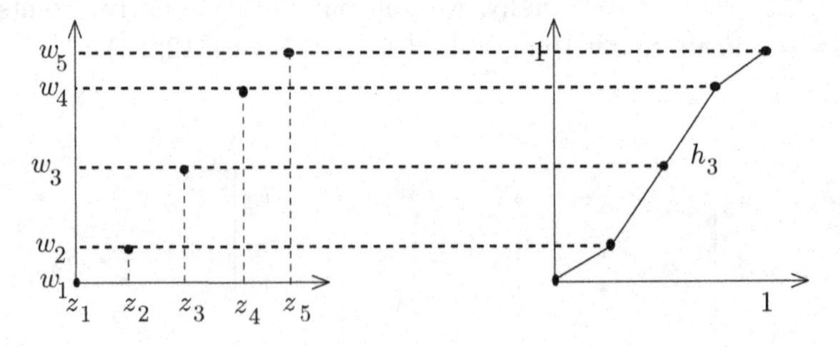


**Figure 12.2.6.** The diagram on the left shows five points on the graph of $h$. The graph of $h_3$, the third piecewise affine approximation to $h$, is on the right.

Thus $h_3$ looks like a better approximation than $h_1$ and $h_2$. In fact, the graph of $h_3$ is starting to follow the characteristic 'S' shape of the graph of $h$ shown in Figure 12.1.3.

To carry the approximation process one step further, we have once again applied Matching Zeroes, this time with $n = 4$, to produce the results shown in Figure 12.2.7. The resemblance, between the graph of the piecewise affine approximation $h_4$ and that of the conjugacy $h$, is now unmistakable.

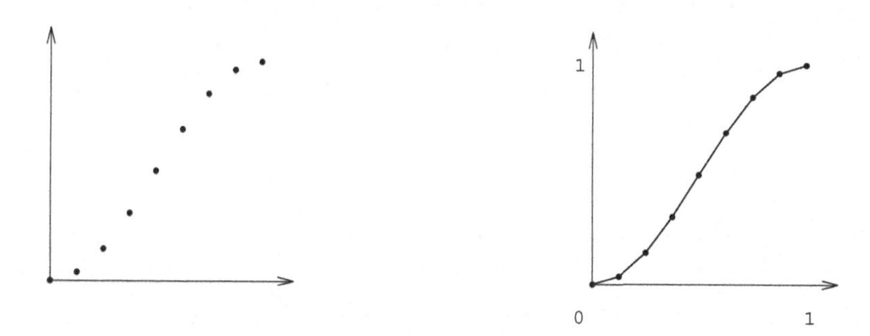


**Figure 12.2.7.** On the left are the points obtained by matching the zeroes of $f^4$ with those of $g^4$. On the right these points are joined by line segments to give the approximation $h_4$ to $h$.

We can proceed like this for as long as we like, taking higher iterates of the tent map and the logistic map and matching their zeroes. We claim that we get more and more points lying closer and closer together on the graph of $h$. When, furthermore, we join these points with line segments we get as close an approximation to the actual graph of $h$ as we wish.

We now show how to justify these claims, not only for the particular case of the tent and the logistic mappings, but for any $f$ and $g$ satisfying suitable conditions.

### Justifying the method: the points $(z_i, w_i)$.

We first justify the claim we have made that the points $(z_i, w_i)$ must lie on the graph of any conjugacy $h$ from $f$ to $g$. This is part (c) of the following lemma.

**12.2.5 Lemma** *Let $f$ and $g$ be one-hump mappings. If there exists a conjugacy $h$ from $f$ to $g$, then:*

*(a) $h$ is an increasing mapping with $h(0) = 0$ and $h(1) = 1$;*

*(b) $h$ maps zeroes of $f^n$ to zeroes of $g^n$ for each $n \in \mathbb{N}$;*

*(c) the points $(z_i, w_i)$ given by the Matching Zeroes method all lie on the graph of $h$.*

*Proof of (a):* Since $h : [0, 1] \to [0, 1]$ is a homeomorphism, and hence invertible, it must be strictly increasing or decreasing.

Suppose $h$ is decreasing. It follows that $h(0) = 1$ and $h(1) = 0$, since $h$ is onto. But $h$ is a conjugacy from $f$ to $g$, and so, in particular,

$$h(f(1)) = g(h(1)).$$

But since $f(1) = 0$ and $h(1) = 0$ this gives

$$h(0) = g(0) = 0$$

and so $h(0) = 0 = h(1)$ contradicting the one-to-oneness of $h$. Hence $h$ cannot be decreasing.

Thus $h$ is increasing and $h(0) = 0$, $h(1) = 1$. ■

*Proof of (b):* Let $z$ be a zero of $f^n$.

To show that $h(z)$ is a zero of $g^n$: use $f^n$ is conjugate to $g^n$.

$$\begin{aligned} g^n(h(z)) &= h(f^n(z)) && \text{by Lemma 12.1.4} \\ &= h(0) && \text{by hypothesis} \\ &= 0 && \text{by Lemma 12.2.5(a).} \end{aligned}$$

Hence $h(z)$ is a zero of $g^n$, for each $n \in \mathbb{N}$. ■

*Proof of (c):* By part (b), $h$ maps the set $\{z_1, z_2, \ldots, z_\ell\}$ of zeroes of $f^n$ to the set $\{w_1, w_2, \ldots, w_\ell\}$ of zeroes of $g^n$. By part (a) of this lemma, $h$ respects the order of these zeroes and hence $h(z_i) = w_i$, for all $i$. Hence the points $(z_i, w_i)$ are all on the graph of $h$. ■

### Justifying the method: the mappings $h_n$

The remaining part of our claim, for the method of Matching Zeroes, is that *the graph of $h_n$ is a good approximation to the graph of $h$* if $n$ is large enough.

Thus we need a way of measuring the closeness of the graph of a mapping $h_n$ to the graph of a mapping $h$. With this end in view, we make the following definition.

**12.2.6 Definition** Let $h : [0,1] \to [0,1]$ and let $\epsilon > 0$.
*The $\epsilon$-strip about $h$* is the set

$$\{(x, y) : x \in [0,1] \text{ and } h(x) - \epsilon < y < h(x) + \epsilon\}$$

■

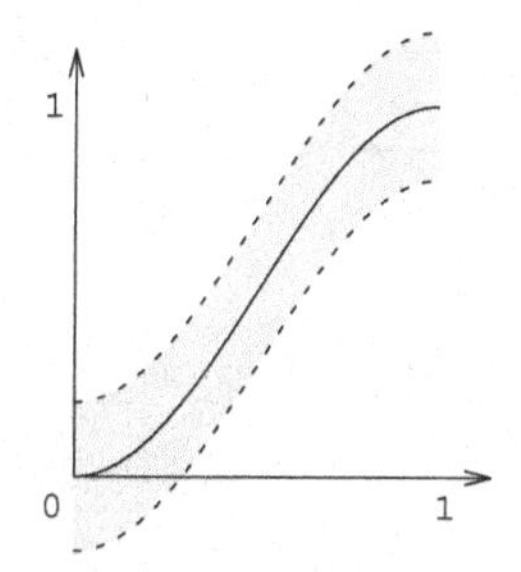


**Figure 12.2.8.** A typical $\epsilon$-strip about $h$. Here $\epsilon = .125$, while $h$ is the conjugacy from the tent mapping $T_4$ to the logistic mapping $Q_4$.

Thus the $\epsilon$-strip about $h$ consists of all the points between the graphs of the mappings $h - \epsilon$ and $h + \epsilon$. We can think of this strip as being close to the graph of $h$ when $\epsilon$ is small. In particular, for practical purposes, the $\epsilon$-strip is indistinguishable from the original graph if $\epsilon$ is less than the width of one pixel when the graph is displayed on the computer screen.

In the following figures, we show at the top of both columns, an $\epsilon$-strip about $h$. In each column, we then test whether the graph of $h_n$ is a subset of the given $\epsilon$-strip.

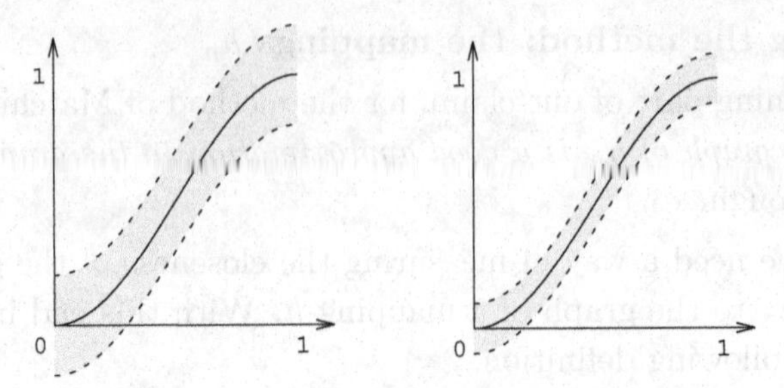


**Figure 12.2.9.** Two $\epsilon$-strips about $h$.

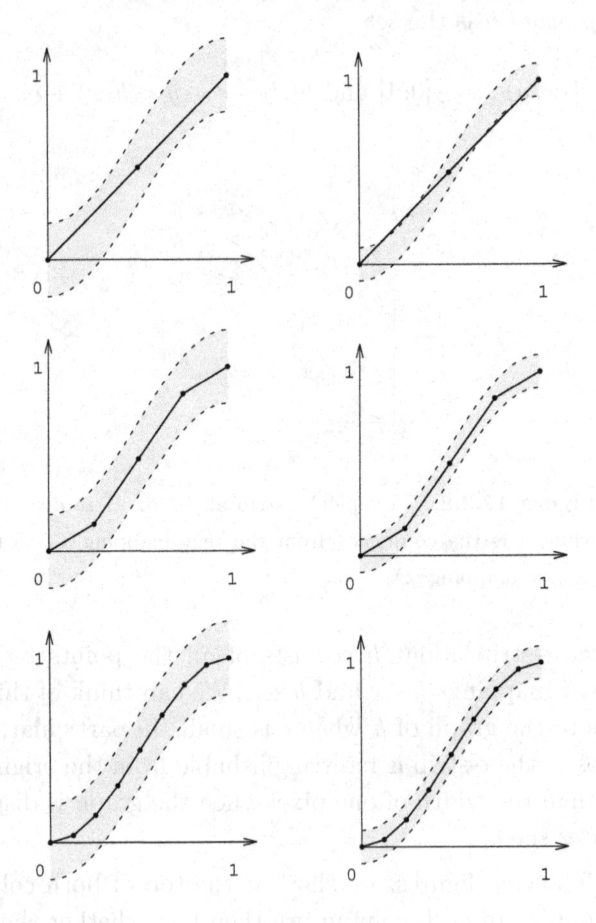


**Figure 12.2.10.** On the left, the graphs of $h_1$, $h_2$ and $h_3$ are all subsets of the $\epsilon$-strip. On the right, however, the graph of $h_1$ fails to be a subset of the $\epsilon$-strip.

The above examples are intended to help you think about the conclusion of the following lemma in terms of graphs.

### 12.2.7 Lemma (Piecewise affine approximation lemma)

*Let $f$ and $g$ be one-hump mappings from $[0,1]$ onto $[0,1]$. Suppose there is a conjugacy $h$ from $f$ to $g$. Let $h_n$ be the piecewise affine mapping obtained from Matching Zeroes. If $e_n$ is the maximum distance between consecutive zeroes of $g^n$ then, for all $x \in [0,1]$,*

$$h(x) - e_n \leq h_n(x) \leq h(x) + e_n$$

*or, more briefly,*

$$h - e_n < h_n < h + e_n.$$

*Proof:* Let $z_i$ and $w_i$ be the $i^{\text{th}}$ zeroes of $f^n$ and $g^n$, as in the method of Matching Zeroes, where $1 \leq i \leq \ell$ and $\ell = 2^{n-1} + 1$.

For $1 \leq i \leq \ell - 1$, we use the notation $[z_i, z_{i+1}] \times [w_i, w_{i+1}]$ to denote the rectangular box shown in Figure 12.2.12.

The two points $(z_i, w_i)$ and $(z_{i+1}, w_{i+1})$ are at opposite corners of this box. The graph of the mapping $h_n$ passes through these two points and runs along the diagonal joining them. By Lemma 12.2.5, the mapping $h$ is also increasing and has a graph that passes through these two points. The fact that $h$ is increasing means that the graph $h$ must also stay inside the box (See Exercise 12.2.4).

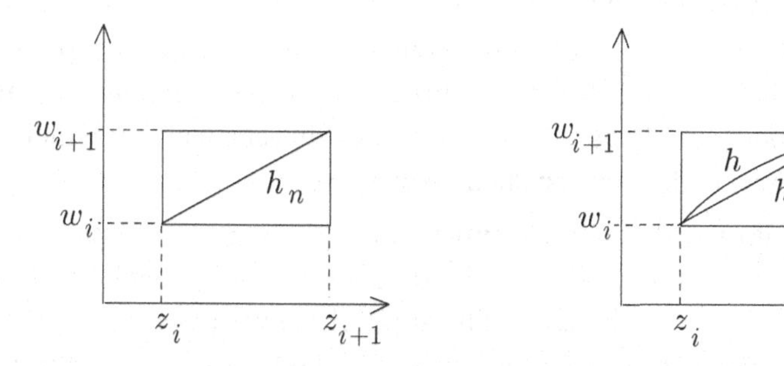


**Figure 12.2.11.** The graphs of both $h$ and $h_n$, restricted to the interval $[z_i, z_{i+1}]$, remain in the box.

But this means that the vertical distance between the graphs of $h$ and $h_n$ cannot exceed the height $w_{i+1} - w_i$ of the box.

Now the vertical distance graphs of $h$ and $h_n$ at a point $x$ is just $|h(x) - h_n(x)|$. Hence for all $x \in [z_i, z_{i+1}]$ where $i$ with $1 \leq i \leq \ell$

$$|h(x) - h_n(x)| \leq w_{i+1} - w_i \leq e_n$$

as $e_n$ is the largest $w_{i+1} - w_i$ and so

$$h(x) - e_n \leq h_n(x) \leq h(x) + e_n$$

as required. ∎

Expressed in terms of graphs, the conclusion of the lemma says

$$\text{graph of } h_n \subseteq e_n\text{-strip about } h. \tag{3}$$

Hence if $e_n$ is small, then the graph of $h_n$ will be close to that of $h$. But if $g$ has wiggly iterates, then, the maximum distance $e_n$ between successive zeroes of $g^n$ is small if $n$ is large. This implies that $h_n$ is a good approximation to $h$ when $n$ is sufficiently large.

For example, if we choose $n$ large enough to ensure that the distance between consecutive zeroes of $g^n$ is less than the width of one pixel, then the graph of $h_n$ would be indistinguishable from that of $h$ itself, when displayed on the screen of a computer.

### Uniform convergence

We shall now discuss the convergence of the sequence of piecewise affine approximations $\{h_n\}_{n=1}^{\infty}$. This is a sequence of *mappings* (rather than numbers), and there is a variety of different ways in which convergence can be defined for such sequences.

In the most widely known type, called *pointwise convergence,* one investigates the convergence of the sequence of values of the mappings at each point of the domain. The type of convergence most relevant to our work on conjugacies, however, is called *uniform convergence.*

Uniform convergence has the advantage that, if a sequence of continuous mappings converges uniformly, then the limit mapping is also continuous. The definition we adopt below is presented in graphical guise so as to foster intuitive understanding.

**12.2.8 Definition** Let $h_n : [0,1] \to \mathbb{R}$ for each $n \in \mathbb{N}$. The sequence of mappings $\{h_n\}_{n=1}^{\infty}$ is said to *converge uniformly* to a mapping $h$ if every $\epsilon$-strip around the mapping $h$ contains the graph of the mapping $h_n$ for all but a finite number of integers $n$. By writing

$$\lim_{n\to\infty} h_n = h \qquad \text{(uniformly)}$$

we mean that the sequence $\{h_n\}_{n=1}^{\infty}$ converges to $h$ uniformly. ∎

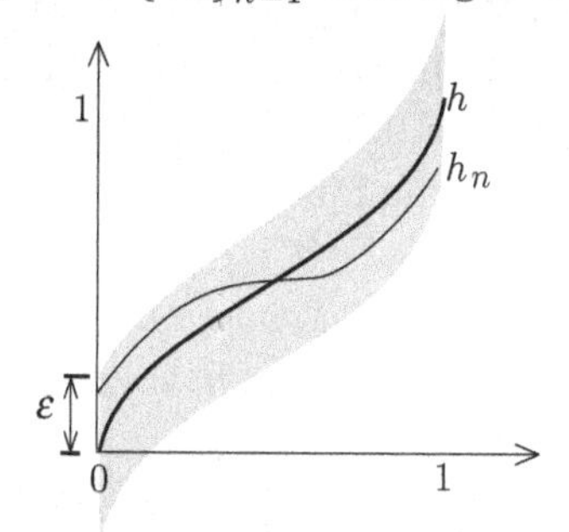


**Figure 12.2.12.** An element $h_n$ of a sequence of mappings inside an $\epsilon$-strip around the limit $h$.

**Squeeze Principle** Suppose that for each $n \in \mathbb{N}$, the graph of the mapping $h_n$ is contained in an $e_n$-strip around the graph of the mapping $h$.

$$\text{If } \lim_{n\to\infty} e_n = 0, \text{ then } \lim_{n\to\infty} h_n = h \text{ (uniformly).}$$

Combining the Piecewise Affine Approximation Lemma with the Squeeze Principle gives the following theorem:

**12.2.9 Theorem** *Let $f$ and $g$ be one-hump mappings and let $g$ have wiggly iterates. Let $\{h_n\}_{n=1}^{\infty}$ be the sequence of piecewise affine mappings produced by Matching Zeroes. If there exists a conjugacy $h$ from $f$ to $g$, then*

$$\lim_{n\to\infty} h_n = h \qquad \textit{(uniformly).}$$

*Proof:* By the Piecewise Affine Approximation Lemma (12.2.7), the graph of $h_n$ is included in the $e_n$-strip around the graph of $h$, where $e_n$ is the largest distance between consecutive of zeroes of $g^n$.
Since $g$ has wiggly iterates, $\lim_{n\to\infty} e_n = 0$. Hence by the Squeeze Principle

$$\lim_{n\to\infty} h_n = h \qquad (uniformly).$$

■

Thus we have expressed the conjugacy $h$ in terms of the piecewise affine approximations, given by Matching Zeroes. But since each $h_n$ is defined unambiguously in terms of the mappings $f$ and $g$, the same is true of $h$. Thus we get the following corollary.

**12.2.10 Corollary** *There is at most one conjugacy $h$ from $f$ to $g$ if they are one hump mappings and $g$ has wiggly iterates.* ■

## Exercises 12.2

12.2.1. Suppose there is a homeomorphism $h : [0,1] \to [0,1]$ under which a one hump mapping $f : [0,1] \to [0,1]$ is conjugate to itself.

(a) Use the method of Matching Zeroes to find each of the piecewise affine approximations $h_1, h_2, h_3, h_4$ to the homeomorphism $h$.

(b) Hence guess the mapping $h$ and then verify that it is a conjugacy from $f$ to $f$.

12.2.2. Let $h : [0,1] \to [0,1]$ be the conjugacy from the tent mapping $T_4$ to the logistic mapping $Q_4$.

(a) Find the graph of $h$ in the text and then deduce the graph of the inverse mapping $h^{-1} : [0,1] \to [0,1]$.

(b) If a point $(z_i, w_i)$ is on the graph of $h$, which point is on the graph of $h^{-1}$?

(c) Sketch the graphs of the inverses of the mappings $h_2, h_3, h_4$ discussed on pages 225-227.

(d) Do your sketches suggest that $h_n^{-1}$ will be a close approximation to $h^{-1}$ when $n$ is large?

(e) Show that for each $n \in \mathbb{N}$, for all $x \in [0,1]$,

$$|h^{-1}(x) - h_n^{-1}(x)| \leq 2^{1-n}.$$

(f) Show that if $n > 4$, then at all points $x \in [0,1]$ the value of $h_n^{-1}$ differs from that of $h^{-1}$ by less than 0.06.

12.2.3. Let $h : [0,1] \to [0,1]$ be a homeomorphism. Prove that $h$ must be either strictly increasing or strictly decreasing.

[Hint. If $h$ is neither strictly increasing nor strictly decreasing, then there exist points $a, b, c \in [0,1]$ such that $a < b < c$ and

$$h(b) > h(a), h(c) \quad \text{or} \quad h(b) < h(a), h(c).$$

In both cases use the Intermediate Value Theorem to derive a contradiction to the assumption that $h$ is a homeomorphism.]

12.2.4. In the proof of Lemma 12.2.7 we stated that 'The fact that $h$ is increasing means that the graph $h$ must also stay inside the box'. Give a careful justification of this assertion.

## 12.3 *EXISTENCE OF A CONJUGACY

In the previous section we investigated Matching Zeroes as a practical method of approximating the graph of a conjugacy $h$ between two one-hump mappings $f$ and $g$ with wiggly iterates.

We showed, furthermore, that the sequence of piecewise affine approximations $\{h_n\}_{n=1}^{\infty}$, constructed in Matching Zeroes, converges uniformly to the conjugacy $h$ *if it exists.* As a corollary of this result we noted that for a given $f$ and $g$ there could be at most one conjugacy, thereby establishing a *uniqueness* result for conjugacies.

We now tackle the problem of *existence:* Given any pair $f$ and $g$ of one-hump mappings with wiggly iterates, is there at least one conjugacy $h$ between them? This is a harder problem than uniqueness because we cannot assume that we know a conjugacy at the start. This places on us the responsibility of showing how to *construct* a conjugacy $h$. As a result, some of the things that were previously called theorems, now become definitions.

### Theorems which become definitions

For example, previously we *proved* the formula $h(z_i) = w_i$, where the zeroes $z_i$ and $w_i$ were matched at the $n$th stage of the approximation process. Now, however, we want to use this formula as part of our *definition* of $h$. To ensure that this definition is unambiguous, we need to know that if, at any stage of the approximation process the zeroes $z_i$ and $w_i$ are matched, then they stay matched at all subsequent stages. This is guaranteed by the following lemma.

**12.3.1 Lemma** *If $z_i$ and $w_i$ are the $i^{\text{th}}$ zeroes of $f^n$ and $g^n$ respectively, then for each $m > n$ they are the $j^{\text{th}}$ zeroes of $f^m$ and $g^m$ for some $j > i$.*

*Proof:* Each time the exponent $n$ is increased by 1, $f^n$ acquires one new zero between each of the old ones, as $f$ is a one-hump mapping. The same is true of $g^n$. Hence the position of each of the zeroes $z_i$ and $w_i$ in the list of zeroes will increase by the same amount. ∎

We do not yet know whether there is a conjugacy $h$ from $f$ to $g$, and so the Piecewise Affine Approximation Lemma cannot be used as it stands. It is easy, however, to modify this lemma so that, instead of referring to a pair of mappings $h$ and $h_n$, it refers to pair of mappings $h_m$ and $h_n$ constructed directly from the mappings $f$ and $g$ as in the method of Matching Zeroes.

### 12.3.2 Lemma (Modified piecewise affine approximation)

*Let $f$ and $g$ be one-hump mappings with wiggly iterates and let $h_n$ be the sequence of piecewise affine mappings produced by applying Matching Zeroes to $f$ and $g$. If $n < m$, then*

$$h_m - e_n < h_n < h_m + e_n.$$

*where $e_n$ is the largest distance between the zeroes of $g^n$.*

*Proof:* Let $z_i$ and $w_i$ be the $i^{\text{th}}$ zeroes of $f^n$ and $g^n$ respectively. Now $h_m$ is an increasing mapping and the points $(z_i, w_i)$ are on the graph of $h_m$ (as well as on the graph of $h_n$), by Lemma 12.3.1. It follows that $h_m$ satisfies the properties of $h$ which were used in the proof of the original Piecewise Affine Approximation Lemma (Lemma 12.2.7). The proof now follows a similar course to that of the original Piecewise Affine Approximation Lemma. ∎

We now need some standard definitions and theorems from analysis[2] which enable us to use our sequence of piecewise affine mappings to obtain a conjugacy. We want to be sure that provided this sequence *of functions* behaves 'in the right way', it will have a limit which is a continuous function. For this purpose, the meaning of 'in the right way' is explained by the next definition.

> **Definition** A sequence of mappings $\{h_n\}_{n=1}^{\infty}$ is called t a *Cauchy sequence* if:
>
> for every $\varepsilon > 0$ there is a number $N$ such that,
>
> for all $m, n \geq N$
>
> there is an $\epsilon$-strip around $h_n$ containing $h_m$.

[2]The further reading section at the end of this chapter discusses some references on the results from analysis used in this section.

Intuitively speaking, a Cauchy sequence of functions is one where the elements are getting closer and closer together as $n \to \infty$ and are desperately trying to converge to some function. The only thing that might prevent a Cauchy sequence from converging, is that the function it is trying to converge to is 'missing' from the set of functions we are considering. Another way of putting this is to say that the set of functions is 'incomplete'.

> **Definition** A set $S$ of functions is said to be *complete* if every Cauchy sequence whose elements are all in $S$ has a limit in $S$.
>
> **Completeness Theorem** The set of continuous mappings from $[0, 1]$ into itself is complete.

Thus, provided we can show that our sequence $\{h_n\}_{n=1}^{\infty}$ is Cauchy, it will have a continuous limit function which will be our candidate for a conjugacy. A further result from analysis makes this easy.

> **Squeeze Principle** A sequence of mappings $\{h_n\}_{n=1}^{\infty}$ is a Cauchy sequence if there is a sequence $\{e_n\}_{n=1}^{\infty}$ of numbers with limit 0 such that whenever $n < m$ the graph of $h_m$ is in the $e_n$-strip around the graph of $h_n$.

Now the sequence $\{e_n\}_{n=1}^{\infty}$ from Lemma 12.3.2 has limit 0. Hence the sequence $\{h_n\}_{n=1}^{\infty}$ is Cauchy, so the above completeness theorem guarantees it has a limit, making it valid to proceed to the following definition.

**12.3.3 Definition** $h = \lim_{n\to\infty} h_n$ where the $h^n$ are the piecewise affine mappings produced by applying Matching Zeroes to $f$ and $g$. ■

It remains to check that $h$ is a homeomorphism and that it satisfies the conjugacy relationship $f \circ h = h \circ g$. These two problems will be tackled separately.

### Showing that $h$ is a homeomorphism

We already know that $h$ is continuous and Exercise 12.3.1 shows that $h$ is onto. We prove that $h$ is one-to-one by contradiction.

Suppose $h$ is not one-to-one, so there are points $a, b \in [0,1]$ such that $h(a) = h(b)$. We may assume that $a < b$. Since the zeroes of $f$ are dense, there are at least two of them, say $y$ and $z$ in the interval $(a, b)$. We may assume that $y < z$.

For some $n$ we must have $f^n(y) = f^n(z) = 0$, so $h_n(y)$ and $h_n(z)$ are distinct zeroes of $g^n$ and hence $h_n(y) < h_n(z)$.

Thus at least one of $h_n(y)$ and $h_n(z)$ is different from $h(a) = h(b)$. We will complete the argument in the case where one of them is strictly greater than $h(a) = h(b)$. The other case is left to Exercise 12.3.2.

Since $h_n$ is increasing, we have $h_n(z) > h(b)$.

Let $\varepsilon = \frac{1}{2}(h_n(z) - h(b))$ so that $\varepsilon > 0$ and $h_n(z) > h(b) + \varepsilon$.

Since Exercise 12.3.1 tells us that $h_m(z) = h_n(z)$ for all $m \geq n$ and each $h_m$ is strictly increasing, we have

$$h_m(b) > h_m(z) = h_n(z) > h(b) + \varepsilon$$

for each $m \geq n$. This is illustrated in Figure 12.3.1.

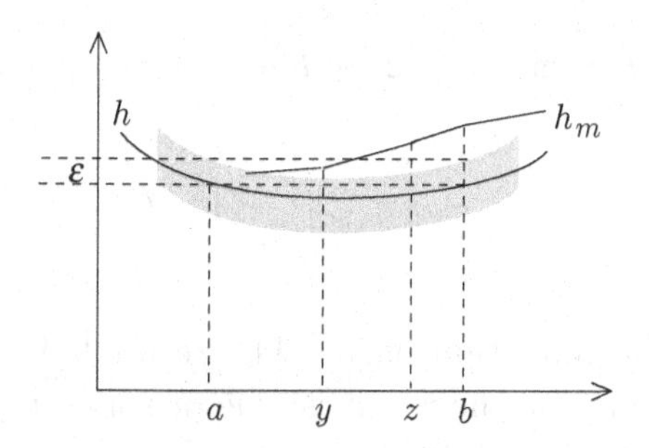


**Figure 12.3.1.** For each $m \geq n$, the increasing map $h_m$ jumps outside the $\epsilon$-strip at $b$.

This contradicts the fact that $h = \lim_{n\to\infty} h_n$ since it means that the graphs of the $h_m$ must lie outside the $\varepsilon$-strip for every $n \geq m$.

The following standard theorem of analysis now completes the proof that $h$ is a homeomorphism. The theorem tells us that the inverse of $h$ is automatically continuous.

> **Theorem** A continuous one-to-one map from a closed bounded interval onto a closed bounded interval is a homeomorphism.

### Showing that $h$ is a conjugacy

Our strategy will be to first show that the conjugacy relationship $f \circ h = h \circ g$ holds on the set of zeroes of iterates of $f$ and then to show that this extends to the whole of $[0, 1]$ by taking limits. To prove that this relationship holds for zeroes of iterates of $f$, we need detailed information about how the $\ell_n (= 2^{n-1} + 1)$ zeroes

$$0 = z_1 < z_2 < \cdots < c < \cdots < z_\ell = 1$$

of $f^n$ are mapped by $f$ (where $c$ is the critical point of $f$). Since zeroes are preserved under iteration, each zero in this list must map to another zero in the list. The following lemma tells us exactly which.

**12.3.4 Lemma** *For each $n \geq 2$ we have $c = z_{\ell_{n-1}}$ and*

$$f(z_i) = \begin{cases} z_{2i-1} & \text{if } i \leq \ell_{n-1} \\ z_{2(\ell_n - i)+1} & \text{if } i > \ell_{n-1} \,. \end{cases}$$

*Proof:* We will use induction on $n$. The result holds for $n = 2$ by Exercise 12.3.3. Suppose the result is true or some $n \geq 2$ and denote the $\ell_{n+1} = 2^n + 1$ zeroes of $f^{n+1}$ by

$$0 = u_1 < u_2 < \cdots < c < \cdots < u_{\ell_{n+1}} = 1 \,.$$

For each $i < \ell_{n-1}$ our inductive hypothesis gives

$$f(z_i) = z_{2i-1} < z_{2i} < z_{2i+1} = f(z_{i+1})$$

so $z_{2i}$ is the *only* zero of $f^n$ in the interval $(f(z_i), f(z_{i+1}))$.

Now $f$ is strictly increasing on $(z_i, z_{i+1})$ since $z_{i+1} \leq c$ by the inductive hypothesis.

Hence the Intermediate Value Theorem yields precisely one $u \in (z_i, z_{i+1})$ such that $f(u) = z_{2i}$. See Figure 12.3.2.

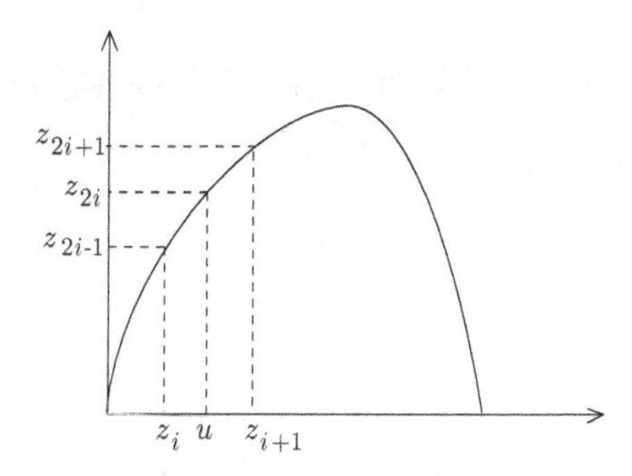


**Figure 12.3.2.** Using the Intermediate Value Theorem to find the 'new' zeroes of $f^{n+1}$.

Thus $u$ is a zero of $f^{n+1}$. In fact it is the only zero of $f^{n+1}$ in $(z_i, z_{i+1})$ because $f^n$ had no zeroes in this interval.

A similar argument for $i > \ell_{n-1}$ and uses the fact that $f$ is strictly decreasing on $(z_i, z_{i+1})$ to show that $f^{n+1}$ has precisely one new zero in this interval. Exercise 12.3.5 gives the details.

It is now clear that the zeroes of $f^n$ and $f^{n+1}$ are related as follows:

$$\begin{array}{ccccccccccc} z_1 & & z_2 & & \dots & z_i & & z_{i+1} & \dots & z_{\ell_n} \\ u_1 & u_2 & u_3 & u_4 & \dots & u_{21-1} & u_{2i} & u_{2i+1} & \dots & u_\ell \end{array}$$

giving $z_i = u_{2i-1}$ for all $i \leq \ell_n$ and hence $u_j = z_{\frac{1}{2}(j+1)}$ when $j$ is odd. We can now use the inductive hypothesis to calculate

$$c = z_{\ell_{n-1}} = u_{2\ell_{n-1}-1} = u_{2(2^{n-2}+1)-1} = u_{2^{n-1}+1} = u_{\ell_n},$$

so the first part of our claim for $f^{n+1}$ holds. It remains to check that the $u_j$ satisfy the formula given in the lemma. There are two cases.

*Case 1:* For $j$ odd, we have $u_j = z_{\frac{1}{2}(j+1)}$ and by Exercise 12.3.4(a) $\frac{1}{2}(j+1) \leq \ell_{n-1} \iff j \leq \ell_n$. Our inductive hypothesis for $f^n$ allows us to apply the formula to $u_j = z_{\frac{1}{2}(j+1)}$. This gives

$$f(u_j) = f(z_{\frac{1}{2}(j+1)}) = \begin{cases} z_j & \text{if } j \leq \ell_n \\ z_{2\ell_n - j} & \text{if } j > \ell_n\,. \end{cases}$$

Now $z_j = u_{2j-1}$. Aplying this fact and Exercise 12.3.4(b) to $z_{2\ell_n - j}$ gives $z_{2\ell_n - j} = u_{2(2\ell_n - j)-1} = u_{2(\ell_{n+1}-j)+1}$, so

$$f(u_j) = \begin{cases} u_{2j-1} & \text{if } j \le \ell_n \\ u_{2(\ell_{n+1}-j)+1} & \text{if } j > \ell_n \end{cases}$$

which says precisely that the formula holds when $j$ is odd.

*Case 2:* For $j$ even and $j < \ell_n$, our constuction above shows that $f(u_j) = z_{2i}$ where

$$z_i = u_{j-1} < u_j < u_{j+1} = z_{i+1}$$

and since $j-1$ is odd we may apply the formula $z_i = u_{2i-1}$ obtained above to the fact that $z_i = w_{j-1}$ to see that $j-1 = 2i-1$ whence $i = \frac{1}{2}j$. Hence $f(u_j) = z_{2i} = z_j = w_{2j-1}$ which shows precisely that the formula holds for even $j < \ell_n$ .

The proof that the formula holds for even $j > \ell_n$ is Exercise 12.3.5.

The result now follows by mathematical induction. ■

**12.3.5 Lemma** *For each $n$ and for each zero $z_i$ of $f^n$,*

$$h \circ f(z_i) = g \circ h(z_i).$$

*Proof:* Let $z_i$ be a zero of $f^n$ for some $n$. Applying Lemma 12.3.4 to $f$ gives

$$h(f(z_i)) = \begin{cases} h(z_{2i-1}) & \text{if } i \le \ell_{n-1} \\ h(z_{2(\ell_n - i)+1}) & \text{if } i > \ell_{n-1} \end{cases} = \begin{cases} w_{2i-1} & \text{if } i \le \ell_{n-1} \\ w_{2(\ell_n - i)+1} & \text{if } i > \ell_{n-1}. \end{cases}$$

But Lemma 12.3.4 can also be applied to $g$ giving

$$g(h(z_i)) = g(w_i) = \begin{cases} w_{2i-1} & \text{if } i \le \ell_{n-1} \\ w_{2(\ell_n - i)+1} & \text{if } i > \ell_{n-1}. \end{cases}$$

Hence $h(f(z_i)) = g(h(z_i))$. ■

Finally, we show that the conjugacy relationship holds for the remaining points in $[0, 1]$ by taking limits.

**12.3.6 Theorem** *If $f$ and $g$ are one-hump mappings with wiggly iterates, then there is a conjugacy $h$ from $f$ to $g$.*

*Proof:* Let $x$ be a point which is not a zero of any iterate of $f$. By Exercise 12.3.6 there is a sequence $\{z_n\}$ of zeroes of iterates of $f$ such that $z_n \to x$ as $n \to \infty$. For all $n$

$$\begin{aligned} h \circ f(z_n) &= g \circ h(z_n) \\ \lim_{n\to\infty} h \circ f(z_n) &= \lim_{n\to\infty} g \circ h(z_n) \\ h \circ f\left(\lim_{n\to\infty} z_n\right) &= g \circ h\left(\lim_{n\to\infty} z_n\right) \qquad \text{by continuity} \\ h \circ f(x) &= g \circ h(x). \end{aligned}$$

■

## Exercises 12.3

Throughout these exercises, let $f$ and $g$ be one-hump mappings with wiggly iterates, let $h_n$ be the sequence of piecewise affine mappings produced by applying the method Matching Zeroes to $f$ and $g$ and let $h = \lim_{n\to\infty} h_n$.

12.3.1. Let $z$ be a zero of $f^n$ for some $n$.

(a) Show that $h_m(z) = h_n(z)$ for all $m \geq n$.

(b) What do you conclude about $h(z)$?

(c) What do you conclude about $h(0)$ and $h(1)$?

(d) Use part (c) and the Intermediate Value Theorem to prove that $h$ is an onto mapping.

12.3.2. Complete the proof that $h$ is a homeomorphism by obtaining a contradiction the case where $h(a) = h(b)$ and $z \in (a, b)$ is a zero of $f^n$ such that $h_n(z) < h(a) = h(b)$.

[Hint: You will need to show that the maps $h_m$ jump out of an $\varepsilon$-strip at $a$ for all $m \geq n$.]

12.3.3. Check that Lemma 12.3.4 is true when $n = 1, 2$ or 3.

[Hint: In the first two cases you can write down the zeroes of $f^n$.]

12.3.4. Check the following properties of $\ell_n = 2^{n-1} + 1$ used in the proof of Lemma 12.3.4.

(a) $\frac{1}{2}(j+1) \leq \ell_{n-1} \iff j \leq \ell_n$;

(b) $2(2\ell_n - j) - 1 = 2(\ell_{n+1} - j) + 1$.

12.3.5. Show that the construction of the 'new' zeroes of $f^{n+1}$ in Lemma 12.3.4 gives

$$f(u_j) = z_{2(\ell_n - i)} \quad \text{where} \quad z_i = u_{j-1} < u_j < u_{j+1} = z_{i+1}$$

for even $j > \ell_n$ and use the result of Exercise 4(b) to show that the formula given in the Lemma holds for such $j$.

12.3.6. Use the definition of a map with wiggly iterates to show that if $x$ is not a zero of any iterate of $f$, there is a sequence $\{z_n\}$ of zeroes of iterates of $f$ such that $z_n \to x$ as $n \to \infty$.

## 12.4 CONJUGACY AND DYNAMICS

If two mappings are mutually conjugate, one can be obtained from the other by a change of coordinates. Hence we might expect them to have essentially the same dynamical behaviour.

To be specific, let $f$, $g$ and $h$ map $[0, 1]$ into itself, and suppose that these two mappings are conjugate via $h$.

We claim that $h$ maps

(a) *orbits of $f$ to orbits of $g$;*

(b) *periodic points of $f$ to periodic points of $g$ with the same prime period;*

(c) *dense orbits of $f$ to dense orbits of $g$;*

(d) *asymptotic orbits of $f$ to asymptotic orbits for $g$ with the same stability;*

and that conjugation respects

(e) *transitivity;*

(f) *chaotic behaviour.*

Some of these results will be proved in the text; others will be set as exercises. In the proofs, we shall repeatedly use the fact that if $f$ is conjugate to $g$ via $h$, then $f^n$ is conjugate to $g^n$ via $h$ for all $n \in \mathbb{N}$; that is,

$$h \circ f^n = g^n \circ h.$$

**12.4.1 Theorem** *If $f$ is conjugate to $g$ via a homeomorphism $h$, then $h$ maps orbits of $f$ to orbits of $g$.*

*Proof:* To show that if $(x_0, x_1, x_2, \dots)$ is an orbit of $f$, then $(y_0, y_1, y_2, \dots)$ is an orbit of $g$, where $y_n = h(x_n)$ for all $n \in \mathbb{N}$.

That is, to show that

$$y_{n+1} = g(y_n)$$

for all $n \in \mathbb{N}$. This is easy because

$$y_{n+1} = h(x_{n+1}) = h(f(x_n)) = g(h(x_n)) = g(y_n).$$

∎

**12.4.2 Theorem** *If $f$ is conjugate to $g$ via a homeomorphism $h$, then $h$ maps every periodic point of $f$ to a periodic point of $g$ with the same period.*

*Proof:* Let $x_0$ be a periodic point of $f$ with period $n$. Hence

$$f^n(x_0) = x_0$$

and so

$$g^n(h(x_0)) = h(f^n(x_0)) = h(x_0).$$

Hence $h(x_0)$ is a periodic point of $g$ with period $n$. ∎

Thus there is a one-to-one correspondence between the periodic points of $f$ and those of $g$, which is given by $x_0 \leftrightarrow h(x_0)$. If $x_0$ has prime period $m$, moreover, then the same is true of the point $h(x_0)$ (see Exercise 12.4.3). Thus the correspondence respects the *prime* periods of the periodic points.

| A homeomorphism $h$ maps every set of points which is dense in its domain to a set of points which is dense in the codomain of $h$. |
|---|

**12.4.3 Theorem** *If $f$ is conjugate to $g$ via a homeomorphism $h$, then a dense orbit for $f$ is mapped by $h$ to a dense orbit for $g$.*

*Proof:* Use the result given in the above box and Theorem 12.4.1. ∎

Since we do not require the homeomorphism $h$ to be differentiable, conjugacy need not preserve differentiability. Hence the multipliers of a fixed point are not necessarily preserved under conjugation. This leads us to make the following definition, in which we capture one of the essential features of an attractor, without using differentiability.

**12.4.4 Definition** A fixed point $a$ of a mapping $f$ is *topologically stable* if there is an interval $I$ containing $a$ such that every orbit under $f$ which starts in $I$ converges to the fixed point $a$. (In particular, every attractor is topologically stable.) ∎

**Continuity and Sequences**

Let $h$ be a continuous mapping from an interval into the real numbers. If $(x_n)_{n=0}^{\infty}$ is a sequence of real numbers such that

$$\lim_{n \to \infty} x_n = a$$

then

$$\lim_{n \to \infty} h(x_n) = h(a).$$

That is, continuous mappings preserve convergence of sequences.

**12.4.5 Theorem** *Let $f$ be conjugate to $g$ via a homeomorphism $h$. If $a$ is a topologically stable fixed point of $f$, then $h(a)$ is a topologically stable fixed point of $g$.*

*Proof:* Since $a$ is a topologically stable fixed point for $f$, there is an interval $I$ containing $a$ such that every orbit under $f$ which starts in $I$ converges to $a$.

Let $J = h(I)$. Exercise 9.5.1 shows that $J$ is an intervaland it contains $h(a)$ since $a \in I$. Also $h(a)$ is a fixed point of $g$ by Theorem 12.4.2.

For any point $y \in J = h(I)$, there is an $x \in I$ such that $y = h(x)$. Since $a$ is a topologically stable and $h$ is continuous

$$\lim_{n \to \infty} f^n(x) = a \implies \lim_{n \to \infty} h(f^n(x)) = h(a)$$

so using the fact that $h$ is a conjugacy

$$\lim_{n \to \infty} h(f^n(x)) = \lim_{n \to \infty} g^n(h(x)) = \lim_{n \to \infty} g^n(y) = h(a).$$

Hence the orbit of $y$ converges to $h(a)$. ∎

**12.4.6 Theorem** *Let $f$ be conjugate to $g$ via a homeomorphism $h$. If $f$ is transitive, then so is $g$.*

*Proof:* Let $f$ be transitive.

To prove that $g$ is transitive; that is, if $I$ and $J$ are open intervals then there is an $n \in \mathbb{N}$ such that $g^n(I) \cap J \neq \varnothing$.

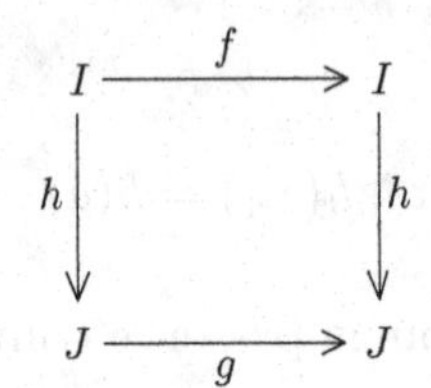


We shall find such an $n$ by using $h$ to pull the problem back onto the domain of $f$, where we can use the transitivity of $f$.

Let $I$ and $J$ be two open intervals. Put $I' = h^{-1}(I)$ and $J' = h^{-1}(J)$. Hence $I'$ and $J'$ are open intervals.
There is an $n \in \mathbb{N}$ such that

$$J' \cap f^n(I') \neq \varnothing.$$

Applying the mapping $h$ to both sides gives

$$h(J') \cap h(f^n(I')) \neq \varnothing$$

and so

$$h(J') \cap g^n(h(I')) \neq \varnothing$$

and hence

$$J \cap g^n(I) \neq \varnothing.$$

Thus $g$ is transitive. ∎

**12.4.7 Theorem** *Let $f$ and $g$ be conjugate via $h$. If $f$ has sensitive dependence, then so does $g$.*

*Proof:* Suppose that $f$ has sensitive dependence, with sensitivity constant $\delta_f > 0$ for $f$.

To show that there is a sensitivity constant $\delta_g > 0$ for $g$.

We ignore quantifiers for the moment. Since $g^n = h \circ f^n \circ h^{-1}$,

$$\begin{aligned} |g^n(x) - g^n(y)| &= |h \circ f^n \circ h^{-1}(x)) - h \circ f^n(\circ(h^{-1}(y))| \\ &= |h(f^n(h^{-1}(x)) - h(f^n h^{-1}(y))| \\ &= |h(f^n(x')) - h(f^n(y'))| \end{aligned}$$

where $x' = h^{-1}(x)$ and $y' = h^{-1}(y)$. Since $f$ is sensitive dependent,we can make $|f^n(x') - f^n(y')| \geq \delta_f$.

Hence Lemma 12.4.8 gives a number $\delta_g$ such that $|h(f^n(x)) - h(f^n(y))| \geq \delta_g$.

Choose $\delta_g > 0$ as in Lemma 12.4.8. To check that $\delta_g$ is a sensitivity constant for $g$:

Let $x \in [0, 1]$. Put $x' = h^{-1}(x)$.

Let $I$ be an interval containing $x$. Put $I' = h^{-1}(I)$,which contains $x'$ and is an interval since $h$ is a homeomorphism.

Since $\delta_f$ is a sensitivity constant for $f$, there is a $y' \in I'$ and an $n \in \mathbb{N}$ such that

$$|f^n(x') - f^n(y')| \geq \delta_f.$$

Hence, by Lemma 12.4.8,

$$|h(f^n(x')) - h(f^n(y'))| \geq \delta_g,$$

and so

$$|h(f^n(h^{-1}(x)) - h(f^n(h^{-1}(y))| \geq \delta_g,$$

That is,

$$|g^n(x) - g^n(y)| \geq \delta_g.$$

Thus $g$ is sensitive dependent with sensitivity constant $\delta_g$. ■

The next proof uses the following result from calculus, first discussed in Section 9.5:

| A function that has as its domain a closed bounded interval and is continuous assumes a minimum value. |
|---|

**12.4.8 Lemma** *Let $h : [0,1] \longrightarrow [0,1]$ be a homeomorphism. For each number $\delta_f > 0$ there is a number $\delta_g > 0$ such that*

$$|v - u| \geq \delta_f \Rightarrow |h(v) - h(u)| \geq \delta_g.$$

*Proof:* Suppose that $|v - u| \geq \delta_f$. Hence either

$$\text{Case A: } v \leq u - \delta_f \qquad \text{or} \qquad \text{Case B: } u + \delta_f \leq v.$$

In case A, since $h$ is monotonic,

$$|h(v) - h(u)| \geq |h(u - \delta_f) - h(u)|.$$

Put $\phi(u) = |h(u - \delta_f) - h(u)|$, to get a mapping $\phi : [\delta_f, 0] \longrightarrow [0,1]$. Since a $h$ is one-to-one mapping, $\phi(u) > 0$. Since $\phi$ is continuous with a closed bounded interval as its domain, it assumes a minimum value. This minimum value, say $\delta_1$, must be positive. Thus there is a number $\delta_1 > 0$ such that $\phi(u) \geq \delta_1$ and hence

$$|h(u - \delta_f) - h(u)| > \delta_1 \qquad \text{for all } u \in [\delta_f, 1] \tag{7}$$

and so

$$|h(v) - h(u)| \geq \delta_1.$$

Similarly, in case B, there is a number $\delta_2 > 0$) such that

$$|h(v) - h(u)| > \delta_2 \qquad \text{for all } u \in [0, 1 - \delta_f]. \tag{8}$$

From (7) and (8) the desired conclusion now follows if we choose $\delta_g = \min(\delta_1, \delta_2)$. ■

## Exercises 12.4

12.4.1. For each step in the proof of Theorem 12.4.1, give the reason which validates the step.

12.4.2. For each step in the proof of Theorem 12.4.2, give the reason which validates the step.

12.4.3. Let $f$ be conjugate to $g$ via a homeomorphism $h$. Let $x_0$ be a periodic point for $f$ with prime period $n$.

(a) Which theorem tells us that $h(x_0)$ is a periodic point for $g$ with period $n$?

(b) Why do the $n$ distinct points of the orbit of $x_0$ under $f$ map to $n$ distinct points? Deduce that $n$ is the *prime* period of $h(x_0)$ under $g$.

12.4.4. Let $f$ be conjugate to $g$ via a homeomorphism $h$. Show that if $f$ has a dense set of periodic points, then so does $g$.

12.4.5. Prove Theorem 12.4.3.

12.4.6. Prove Theorem 12.4.5.

12.4.7. For each step in Theorem 12.4.6, give the reason which validates that step.

---

## Additional reading for Chapter 12

In [Del], topological conjugacy is introduced in Section 1.7 just before chaos is defined together with all its properties. This book does not cover the concept of change of coordinates.

Section 5.11 of [PMJPSY] contains very detailed mathematics behind the equivalence between the tent map and the logistic map. The equivalence established by the nonlinear change of coordinate shows how it preserves the properties between the two maps.

Section 10.6 of [PJS] first covers the discussion given in [PMJPSY]. This motivates the introduction of conjugacy and its properties in Section 10.7. As an example, an affine linear transformation is then constructed for the conjugacy between quadratic polynomials.

Equivalent Dynamical Systems are introduced briefly in Section 5 of Chapter 4 of [Ba] without being much related to the change of coordinates described earlier in Chapter 3.

Cauchy sequences, completeness and theorems connecting them are discussed in just about every analysis texbook. See chapters 3 and 7 of [Di] for example. Good treatments are also given in Chapter 2 of [Ba] and in Chapter 2 of [Ed].

---

## References

[Ba] Michael F. Barnsley, *Fractals Everywhere,* 2nd edition, Academic Press Professional, Boston 1993.

[De1] Robert L. Devaney, *An Introduction to Chaotic Dynamical Systems: Second Edition,* Addison-Wesley, Menlo Park California, 1989.

[Di] J. Dieudonné, *Foundations of Modern Analysis,* Academic Press, London, 1969.

[Ed] Gerald Edgar, *Measure, Topology and Fractal Geometry,* Springer-Verlag, New York, 1990.

[PJS] Hans-Otto Peitgen, Hartmut Jürgens and Dietmar Saupe, *Chaos and Fractals, New Frontiers of Science*, Springer-Verlag, New York, 1992.

[PMJPSY] Hans-Otto Peitgen, Evan Maletsky, Hartmut Jürgens, Terry Perciante, Dietmar Saupe, and Lee Yunker, *Fractals for the Classroom: Strategic Activities Volume Two,* Springer-Verlag, 1992. (two volumes)

# 13

# WIGGLY ITERATES, CANTOR SETS AND CHAOS

*Our study of chaos so far has been restricted to 'chaos on the interval* $[0, 1]$*'; that is, to the chaotic behaviour of functions which map the closed interval* $[0, 1]$ *onto itself. We have ignored what happens to other types of mappings, such as those in the tent and logistic families for parameter values* $\mu > 4$.

*Although these mappings do not map the interval* $[0, 1]$ *into itself, there are many subsets of* $[0, 1]$ *which they do map into themselves. Such sets are called* invariant *sets for the mapping. We show that each mapping, in a certain class, has a largest invariant set, which we denote by* $C$.

*To study the dynamics of the mapping on* $C$, *we use the concept of 'wiggly iterates', which enables us to develop the theory in close analogy with that developed earlier for mappings from* $[0, 1]$ *onto* $[0, 1]$.

*There are two cases which can occur:*
*(i)* $f$ *has wiggly iterates and the invariant set is an esoteric type of set known as a Cantor set. We can show in this case that the mapping, when restricted to the Cantor set, has chaotic behaviour in the sense of Devaney.*
*(ii)* $f$ *does not have wiggly iterates. The invariant set is not a Cantor set and the behaviour of the mapping will not usually be chaotic on the invariant set.*

## 13.1 ITERATING ONE-HUMP MAPPINGS

We now consider mappings like

*the logistic mapping* $Q_\mu(x) = \mu x(1-x)$ and

*the tent mapping* $T_\mu(x) = \frac{\mu}{4}(1 - |2x-1|)$

when $\mu > 4$. For these values of the parameter, the graphs of $Q_\mu$ and $T_\mu$ 'break out' of the box $[0,1] \times [0,1]$, as in Figure 13.1.1. Hence it is no longer true that $Q_\mu$ and $T_\mu$ map the interval $[0,1]$ into itself. As a consequence, we have now to work out the domains of the iterates, as well as their values.

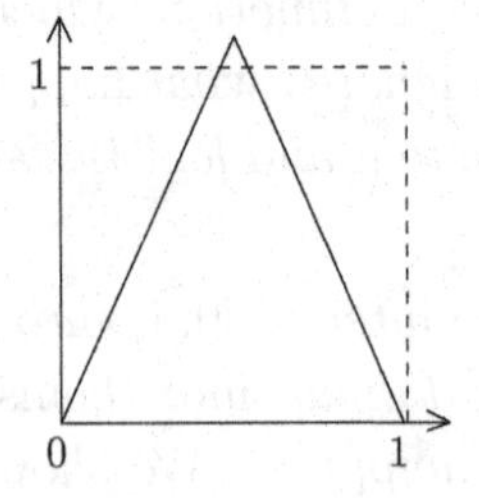


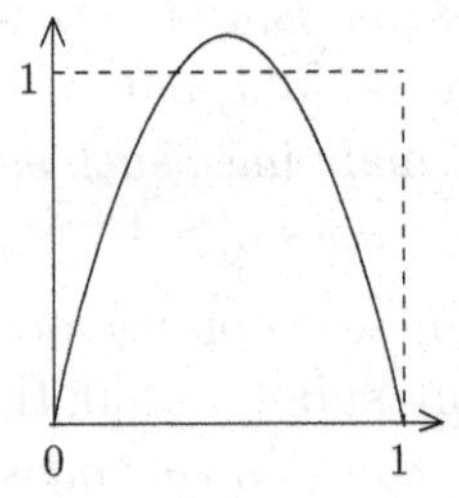


**Figure 13.1.1** $T_\mu$ and $Q_\mu$ when $\mu = 4.2$.

### Tent mapping

We use one of the tent mappings to illustrate how to work out the domains of the iterates. In the case $\mu = 6$ the tent mapping $T = T_\mu$ can be written as

$$T(x) = \begin{cases} 3x & \text{if } x \le \frac{1}{2} \\ 3 - 3x & \text{if } x > \frac{1}{2}. \end{cases}$$

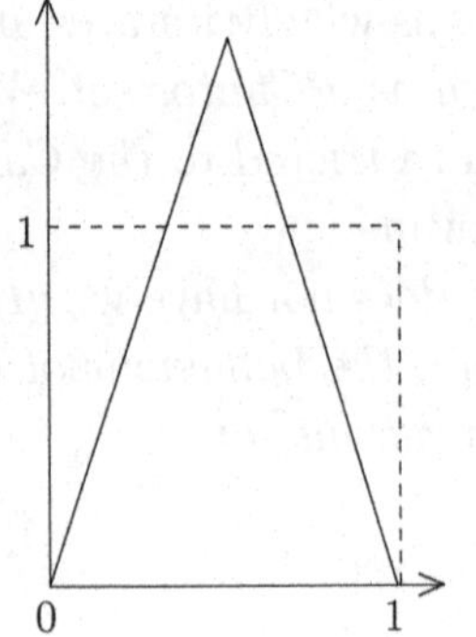


**Figure 13.1.2.** The tent map acting on the unit interval.

### Finding domains

The range of $T$ is larger than its domain. Hence to define the iterates of $T$, we now use the extended definition of composition.[1]

Roughly speaking this is the largest set of $x$ for which the formula $f \circ g(x) = f(g(x))$ makes sense.

The domain of $f \circ g$ can be strictly smaller than the domain of $g$, giving rise to a progressive loss of domain under iteration. With this extended definition, the associative law for composition still holds. As a consequence, powers still commute: $f^n \circ f^m = f^m \circ f^n$.

The domain of $T^1 = T$ is $[0, 1]$, while the domain of $T^2$ is given by

$$\begin{aligned}\text{dom } T^2 &= \text{ dom } T \circ T \\ &= \{x : x \in \text{ dom } T \text{ and } T(x) \in \text{ dom } T\} \\ &= \{x : x \in [0, 1] \text{ and } T(x) \in [0, 1]\}.\end{aligned}$$

Thus to get the domain of $T^2$ we remove from the interval $[0, 1]$ the points which $T$ maps outside the interval $[0, 1]$ on the vertical axis. This amounts to removal of the open middle third of the interval $[0, 1]$. The resulting domain for $T^2$ is shown in Figure 13.1.3, just beneath the graph of $T$ on the left.

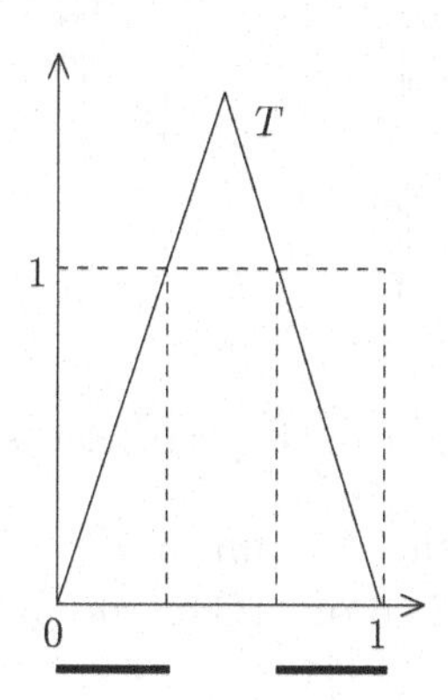


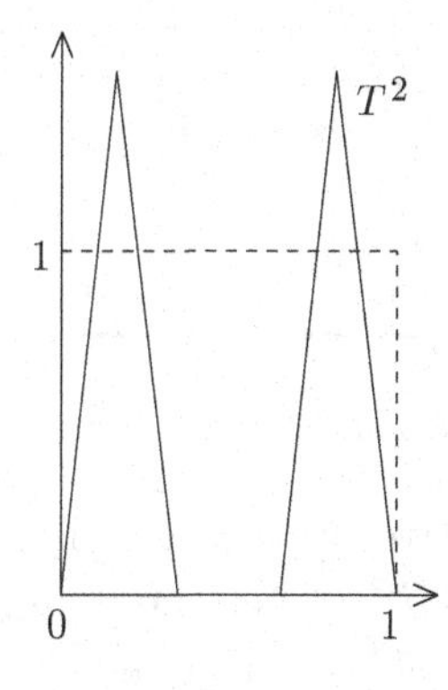


**Figure 13.1.3.** The set of points which $T$ maps into $[0, 1]$ is the domain of $T^2$.

Having found the domain, we can now get the graph of $T^2$ by using

[1]Here we define the domain of $f \circ g$ as the set $\{x \in \text{dom } g : g(x) \in \text{dom } f\}$ (provided this set is not empty).

graphical composition. This gives the result shown in Figure 13.1.3, on the right.

Next, the domain of $T^3$ is given by

$$\begin{aligned} \text{dom } T^3 &= \text{dom } T \circ T^2 \\ &= \{x : x \in \text{dom } T^2 \text{ and } T^2(x) \in \text{ dom } T\} \\ &= \{x : x \in \text{dom } T^2 \text{ and } T^2(x) \in [0,1]\}. \end{aligned}$$

Thus the domain of $T^3$ is obtained from the domain of $T^2$ by removing the points which $T^2$ maps outside $[0, 1]$.

The domain of $T^3$ is shown in Figure 13.1.4, on the left just beneath the graph of $T^2$. It is obtained by removing the open middle third of each of the two intervals which make up the domain of $T^2$. We can then get the graph of $T^3$ by using graphical composition. The result is shown in Figure 13.1.4, on the right.

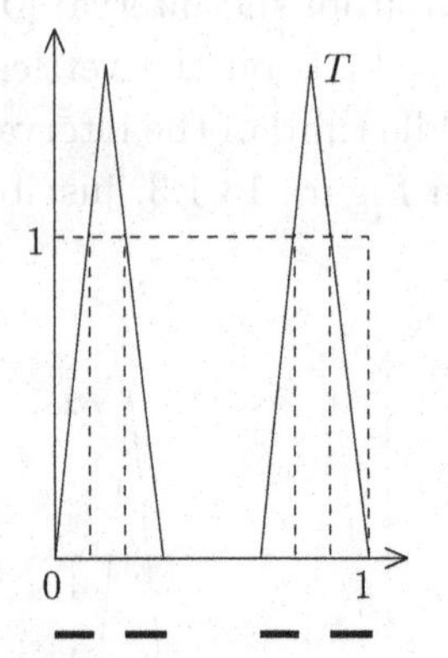


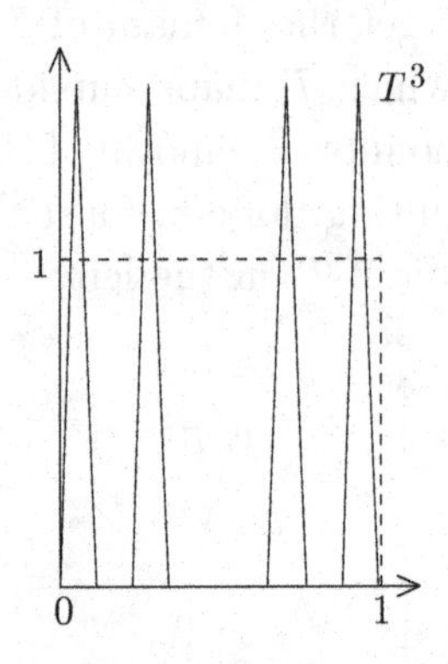


**Figure 13.1.4.** The set of points which $T^2$ maps into $[0, 1]$ is the domain of $T^3$.

Thus we use the graph of $T$ to find the domain of $T^2$; then the graph of $T^2$ to get the domain of $T^3$; and so on. (Can you guess the zeroes of $f^3$ at this stage?)

### Logistic mapping

A similar type of iterate is obtained by iterating a member of the logistic family, say, $Q = Q_\mu$ with $\mu > 4$. Once again, for each iterate we must remove the part of the domain which maps outside the interval $[0, 1]$. After this has been done we can form the next iterate. Graphs of successive iterates are shown in Figure 13.1.5. Beneath the graph of each iterate is shown the domain of that iterate.

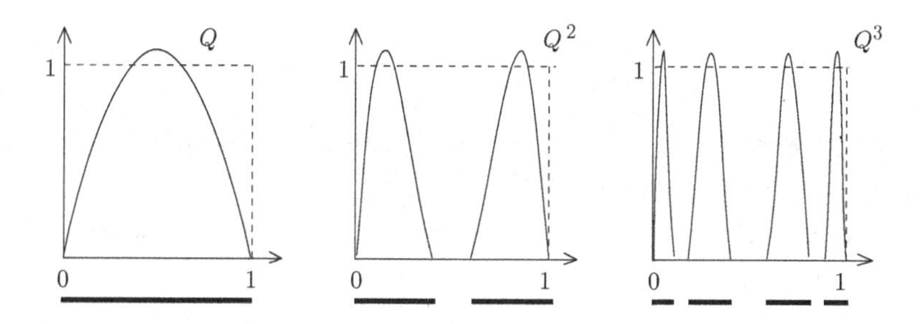


**Figure 13.1.5** Some iterates of the logistic mapping $Q = Q_\mu$ for $\mu = 4.2$.

## One-hump mappings

The above examples suggest the following definitions for a continuous mapping $f$ from $[0, 1]$ onto $[0, r]$ where $r > 1$.

The mapping $Q$ has a single hump, hence:

**13.1.1 Definition** Let $r > 1$ and $a < b < c$. We say that a continuous mapping $f : [0, 1] \to [0, r]$ is a *one-hump mapping on* $[a, c]$ if it strictly increases from $f(a) = 0$ to $f(b) = r > 1$, and then strictly decreases to $f(c) = 0$. ∎

The mapping $Q^2$ has a pair of humps, their bases being separated by an interval. Again, $Q^3$ has four humps, their bases being separated by three intervals, and so on. We introduce the following definition to describe such situations.

**13.1.2 Definition** Let $m \in \mathbb{N}$. We call $f^n$ an *m-hump mapping on* $[0, 1]$ if there are $2m$ points,

$$0 = x_0 < x_1 < \cdots < x_{2m-1} = 1$$

such that $f^n$ is a one-hump mapping on each of the $m$ closed intervals

$$[x_0, x_1], [x_2, x_3], [x_4, x_5], \ldots, [x_{2m-2}, x_{2m-1}].$$

The $i$th of these intervals is called the *base* of the $i$th hump of $f^n$. ∎

The remaining $m-1$ open intervals, which fill the *gaps* between the above intervals, are

$$(x_1, x_2), (x_3, x_4), \ldots, (x_{2m-3}, x_{2m-2}).$$

Since $f^n$ assumes the value 0 at the points $x_0, x_1, \ldots, x_{2m-1}$, these points are called the *zeroes* of $f^n$.

The bases of the humps are the *maximal*[2] closed intervals which are subsets of the domain of $f^n$.

The definitions are illustrated in Figure 13.1.6, which shows a two-hump mapping.

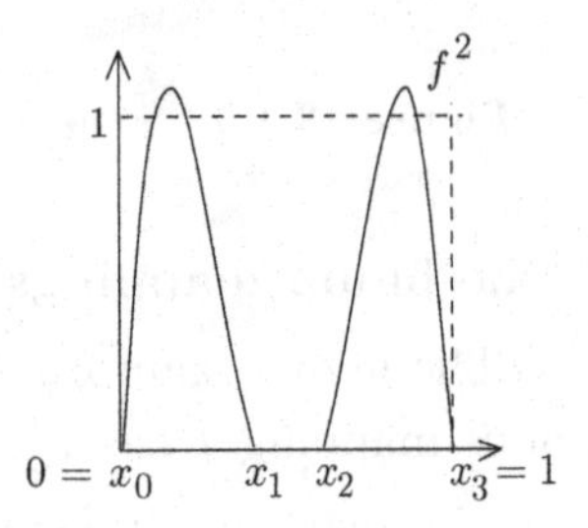


**Figure 13.1.6.** Here $f^2$ has two humps. The base of the first hump is $[x_0, x_1]$ and the base of the second hump is $[x_2, x_3]$. The zeroes of $f^2$ are the numbers $x_0, x_1, x_2, x_3$. The interval $(x_1, x_2)$ fills the gap between the bases of the humps.

The following lemma is suggested by the graphs in Figure 13.1.6.

**13.1.3 Lemma** *Let $r > 1$. If $f : [0,1] \to [0,r]$ is a one-hump mapping, then $f^n$ is a $2^{n-1}$-hump mapping. The domain of $f^n$ is the union of the bases of its humps.*

*The humps of $f^n$ and $f^{n+1}$ are related as follows: an open interval is removed from the base of a hump of $f^n$; this produces a pair of closed intervals which are bases of humps for $f^{n+1}$.*

*Proof:* This is left as a project. Follow the proof of Lemma 7.2.5, making any necessary changes. ∎

## Another approach

Our starting point in this section was a mapping from $[0,1]$ into $[0,r]$. It is more usual, however, to start from a mapping from $\mathbb{R}$ into $\mathbb{R}$.

---

[2]The word *maximal* means that there are no strictly larger intervals which satisfy the given condition (being closed intervals in the domain of $f^n$). A *larger* interval is an interval which includes the given interval.

Thus, for example, the formula giving the values of the logistic mapping $Q = Q_6$ can be used for all $x \in \mathbb{R}$ (and not merely for $x \in [0,1]$). Hence the graph of $Q$ shown in Figure 13.1.1 gets extended to the one shown in Figure 13.1.7, on the left.

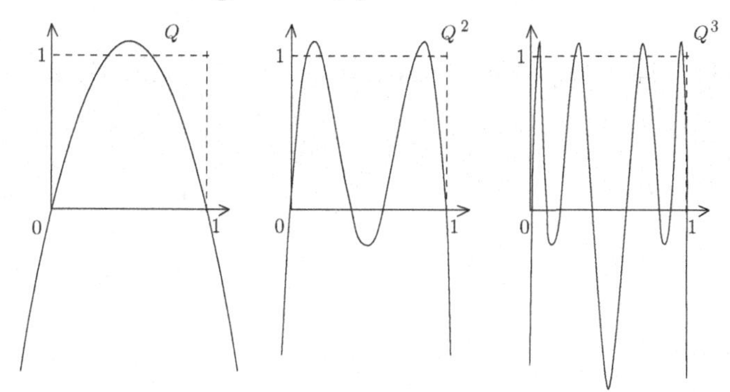


**Fig.13.1.7.** Some iterates of the Logistic mapping $Q = Q_{4.2}$.

This approach avoids the loss of domain under iteration. What happens instead is that if some iterate of a point $x_0$ is negative, then the sequence of iterates of $x_0$ approaches $-\infty$. We prefer our approach because it leads to pictures which are more suggestive of wiggly iterates.

## Return to the tent mapping

The tent mapping, being a one-hump mapping, satisfies the hypothesis and hence also the conclusion of Lemma 13.1.3. Thus the graph of the $n$th iterate of the tent mapping must consist of $2^{n-1}$ humps.

Because the tent mapping is very simple, we can derive some further interesting details about its iterates. For example, we can show that the humps in the iterates of the tent mapping all form *isosceles triangles.* We can prove such things by mathematical induction. The following lemma carries through the inductive step.

**13.1.4 Lemma** *As in Figure 13.1.8, let $H$ and $T$ map the two intervals $I \subseteq [0,1]$ and $[0,1]$ onto the same interval $[0,r]$ where $r > 1$.*

*If the graphs of $T$ and $H$ form isosceles triangles of height $r$, then the graph of $T \circ H$ forms a pair of isosceles triangles also of height $r$.*

*Each of the four triangular humps maps the middle fraction $\alpha$ of its base outside the interval $[0, 1]$, where $\alpha = 1 - 1/r$ . Hence the domain of $T \circ H$ consists of the interval $I$ with the middle fraction $\alpha$ removed.*

*Proof:* This is a simple application of graphical composition. ∎

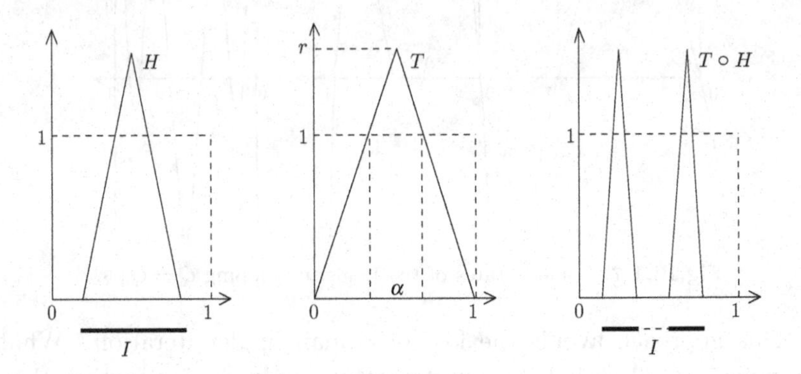


**Fig. 13.1.8.** Composite of two tent mappings is a pair of tent mappings.

We can now prove what Figures 13.1.3 and 13.1.4 suggest about the graph of $T^n$. The results we want to prove are stated in the following theorem, which we shall prove by induction. The identity $T^{n+1} = T \circ T^n$ will be used to reduce the inductive step in the proof to the above lemma. To make the proof more transparent, we deal only with the special case $T = T_6$, so that $r = 3/2$ and $\alpha = 1/3$.

**13.1.5 Theorem** *Let $T = T_6$. The graph of $T^n$ consists of $2^{n-1}$ isosceles triangles, each of height $3/2$ and base length $(1/3)^{n-1}$.*

*The domain of $T^n$ can be obtained recursively from the results*

*(a) dom $T^1 = [0, 1]$;*

*(b) dom $T^{k+1}$ is the result of removing the open middle thirds of all the maximal closed intervals in dom $T^k$.*

*Proof:* The result claimed for the graph of $T^n$ is true when $n = 1$. Let $n \geq 1$ be an integer for which the result is true. Consider one of the humps (isosceles triangles) $H$, with an interval $I$ as domain, in the graph of $T^n$. Since composition behaves naturally with respect to restriction of domain,

$$T^{n+1}|I = (T \circ T^n)|I = T \circ (T^n|I) = T \circ H$$

Lemma 13.1.4 now shows that the graph of $T \circ H$ consists of *two* isosceles triangles, each having *one third of the length* of those in $T^n$.

Thus the total number of triangles has doubled, from $2^{n-1}$ to $2 \times 2^{n-1} = 2^n$, while the base length of the triangles has shrunk by a factor of one third, from $(1/3)^{n-1}$ to $(1/3) \times (1/3)^{n-1} = (1/3)^n$. Hence the result claimed for the graph is true when $n$ is replaced by $n + 1$. The claim is therefore true, for all $n$, by the principle of mathematical induction. ■

## Exercises 13.1

13.1.1.

(a) Give examples of three different open intervals which are subsets of the closed interval $[0, 1]$. Is there a maximal open interval with these properties?

(b) Is there a maximal open interval which is a subset of the open interval $(0, 1)$? If so, which interval is it?

(c) Does the set $[0, 1] \cup [2, 3]$ have any maximal closed subintervals? If so, state which intervals they are.

(d) Does the set $[0, 1] \cup [2, 3]$ include a *largest* closed interval?

13.1.2.

(a) Let $T = T_6$. How many maximal closed intervals are there in the domains of each of the following mappings: $T^1$, $T^2$, $T^3$ ?

(b) Find the endpoints of these intervals and hence write each domain as a union of disjoint closed intervals.

13.1.3. Use the fact that $T^4 = T^3 \circ T$ to show that

$$\text{dom } T^4 = \{x : x \in [0, 1] \text{ and } T(x) \in \text{dom } T^3\}.$$

13.1.4. Use the fact that $T^4 = T \circ T^3$ and the definition of domain for a composite to show that

$$\text{dom } T^4 = \{x : x \in \text{dom } T^3 \text{ and } T^3(x) \in [0,1]\}.$$

Explain how you would use this result to find the domain of $T^4$ from Figure 13.1.4.

13.1.5. For $n \geq 1$ use the formulas $T^n = T \circ T^{n-1} = T^{n-1} \circ T$ to get two different expressions for the domain of $T^n$.

13.1.6. Refer to the graph of $Q^3$ shown in Figure 13.1.5.

(a) How many humps does $Q^3$ have? How many zeroes?

(b) On a copy of the graph of $Q^3$ show the zeroes $x_0, x_1, \ldots, x_7$. Express the bases of the humps in terms of these zeroes. In the domain of $Q^3$, which are the maximal closed intervals ?

13.1.7. Let $Q$ be as in Figure 13.1.5. What does Lemma 13.1.3. predict about the number of humps for the mapping $Q^4$?

13.1.8. If the domain of the $n$th iterate of the tent mapping $T = T_6$ is expressed as a union of maximal closed intervals,

(a) how many closed sets are there in the union?

(b) what is the length of each of these intervals?

(c) what happens to these lengths as $n \to \infty$?

13.1.9. State a more general version of Theorem 13.1.5 which is applicable to a tent mapping $T = T_\mu$ of arbitrary height $r > 1$ and with the fraction of each interval in the domain removed at each iteration equal to $\alpha = 1 - 1/r$.

13.1.10. Sketch the graph of $T_\mu$ when $\mu = 8$ and state the values of the parameters $r$ and $\alpha$ for this mapping. Sketch the graphs of $T_\mu^2$ and $T_\mu^3$ (using the answer to Exercise 9). What fraction of each interval is removed from the domain under one iteration?

13.1.11. Find $\mu$ such that, under iteration by the tent map $T_\mu$, middle fifths of existing intervals are removed from the domain.

## 13.2 CONSTRUCTING AN INVARIANT SET

Throughout this section, $r > 1$ and

$$f : [0, 1] \to [0, r]$$

denotes a one-hump mapping.

Each time we iterated a mapping of this type, there was a loss of domain for the iterate. *Does this process ever settle down so that eventually there is no further loss of domain?* To answer this question we introduce the idea of an invariant set for a mapping.

**13.2.1 Definition** We say that a subset $C$ of the domain of a mapping $f$ is an *invariant set* for $f$ if

$$f(C) \subseteq C.$$

■

**13.2.2 Example** *Let $f$ be a mapping with an orbit of prime period 2. Let $x_1$ and $x_2$ be the two distinct points of the orbit. Show that the set $S = \{x_0, x_1\}$ is an invariant set for $f$.*

*Solution:*

$$\begin{aligned} f(S) &= \{f(x) : x \in S\} \\ &= \{f(x_0), f(x_1)\} \\ &= \{x_1, x_0\} \\ &= \{x_0, x_1\} \\ &= S. \end{aligned}$$

Thus $f(S) \subseteq S$) and so $S$ is an invariant set for $f$. ■

The mapping $f|C$, obtained by restricting the domain of $f$ to an invariant set $C$, has a subset of $C$ as its range. This is indicated by writing $f|C : C \to C$, or more simply just

$$f : C \to C.$$

Thus our original question will be answered affirmatively if, for each one-hump mapping $f$, we can find an invariant set $C$.

To aid the search for an invariant set, we now take a closer look at the domains of the iterates of a typical mapping $f$ of the type we are discussing.

The following lemma, which is just a restatement of Lemma 13.1.3, is the basic result on which we build. Here, as in the rest of this section, *$C_n$ is the domain of $f^n$ where $f$ is a one-hump mapping.*

**13.2.3 Lemma** *The domains of the iterates satisfy the conditions:*

(a) *$C_1 = [0, 1]$ and, for $n = 1, 2, \dots$ ;*

*$C_n$ is the union of $2^{n-1}$ closed disjoint intervals; and*

(b) *$C_{n+1}$ is obtained from $C_n$ by removing an open interval from inside[3] each maximal closed interval in $C_n$, This leaves two closed intervals in $C_{n+1}$ in place of each closed interval in $C_n$.*

A typical pattern for the domains of the iterates of a one-hump mapping is shown in Figure 13.2.1. It is easy to check that the pattern conforms with Lemma 13.2.3 since $C_{n+1}$ is obtained from $C_n$ by deleting the open middle thirds of all its maximal closed intervals.

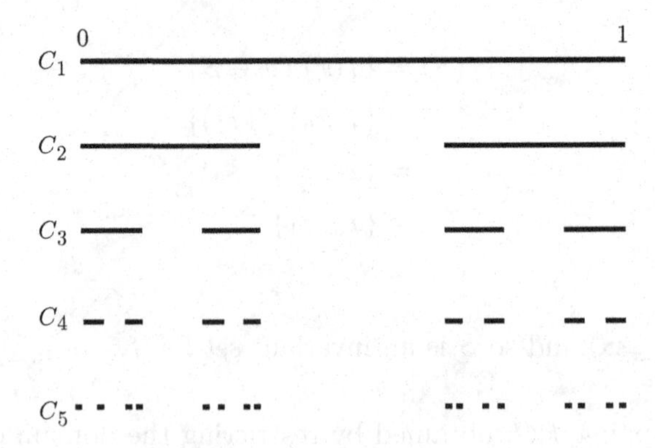


**Figure 13.2.1.** Domains of iterates $T^1$, $T^2$, $T^3$, $T^4$, $T^5$ of the tent mapping.

[3] We say that an open interval $(c, d)$ is *inside* a closed interval $[a, b]$ if removing $(c, d)$ from $[a, b]$ leaves a pair of closed intervals. This means that $a < c$ and $d < b$.

To get an intuitive grasp of what is going on in Figure 13.2.1, it may be helpful to think of the sets $C_1, C_2, C_3, \ldots$ as successive approximations to some 'limiting' set. As $n$ approaches $\infty$, the set $C_n$ approaches this limit.[4]

*Which set should the limit be?* Well, its what's left at the 'end' of the process, after all the middle thirds have been removed. It consists of all the points which the sets $C_1, C_2, C_3, \ldots$ have in common; that is, the set

$$C = \bigcap_{n=1}^{\infty} C_n.$$

This notation will be retained throughout this section.

If we are prepared to think of $C_{n+1}$ and $C_n$ as being close to the 'limit' $C$, then the following lemma suggests $C$ will be an invariant set for $f$.

**13.2.4 Lemma** *If $f$ is a one-hump mapping and if $C_n$ denotes the domain of $f^n$ for each $n \in \mathbb{N}$, then*

$$f(C_{n+1}) \subseteq C_n.$$

*Proof:* See Exercise 13.2.3. ∎

On the basis of this lemma, it is now possible to get the result we have been aiming for.

**13.2.5 Theorem** *If $f$ is a one hump mapping, then $f(C) \subseteq C$ and hence $C$ is an invariant set for $f$.*

*Proof:* See Exercise 13.2.4. ∎

**13.2.6 Theorem** *If $f$ is a one-hump mapping, then $f(C) = C$ and hence $f : C \to C$ is onto.*

*Proof:* See Exercise 13.2.8. ∎

We can prove a stronger result, just as easily, which tells us that the base of every hump contains a smaller copy of the whole invariant set. This is called the *fractal* property of the invariant set $C$.

---

[4]It is possible to make formal sense of this intuition by making the subsets of $[0, 1]$ into a metric space. This involves using the *Hausdorff metric*, as explained, for example, in Barnsley's book on Fractals.

**13.2.7 Theorem (Fractal Theorem)** *If $f$ is a one-hump mapping then each hump of $f^n$ maps the set 'base of hump intersected with C' onto $C$.*

*Proof:* See Exercise 13.2.10. ■

This theorem is illustrated in Figure 13.2.2, where we have chosen $f$ to be the logistic mapping $Q_{4.2}$ and $n$ to be 2.

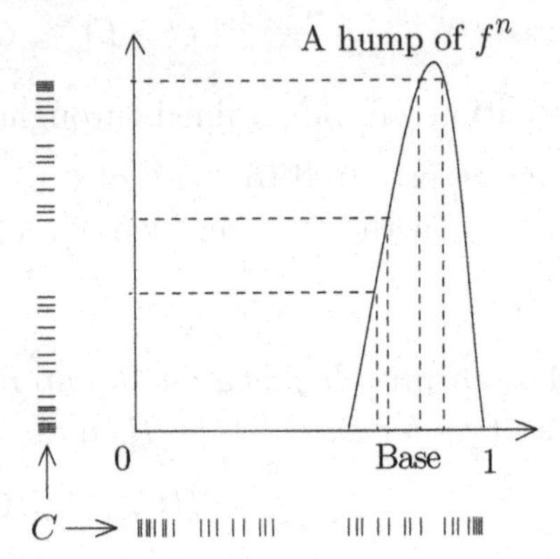


**Fig.13.2.2.** Fractal property of the invariant set $C$. One copy of $C$ is placed along the domain axis, the other along the codomain axis.

The base of a hump of $f^n$ intersects the set $C$, giving a smaller part of $C$. The hump then maps this part back onto the whole of $C$.

Thus the set $C$ contains smaller 'copies' of itself. The smaller we choose the hump, the smaller will be the copy.

Now that we have an invariant set, it would be nice to know that it is reasonably large, or at least that it is non-empty. The following theorem tells us that the invariant set $C$ we have lighted on happens to be the *largest* invariant set of the mapping $f$.

**13.2.8 Theorem** *If $f$ is a one-hump mapping, then every invariant set for $f$ is a subset of $C$.*

*Proof:* Let $S \subseteq [0, 1]$ be an invariant set for $f$. Let $x \in S$.

Since $S$ is an invariant set, $f^n(x) \in S$ for every $n \in \mathbb{N}$. Hence $x \in [0, 1]$ and $f^n(x) \in [0, 1]$ for every $n \in \mathbb{N}$.

From Exercise 13.2.7 it follows inductively that $x \in C_n$ if $n \in \mathbb{N}$. Hence $x \in C$. Thus $S \subseteq C$. ■

That the set $C$ is the *largest* invariant subset of the mapping $f$ will now be used as a neat technique for showing certain sets are subsets of $C$.

**13.2.9 Corollary** *The set of periodic points of $f : [0,1] \to [0,r]$ is a subset of $C$.*

*Proof:* The set of all periodic points of $f$ is invariant under $f$ (see Exercise 13.2.11). Hence by Theorem 13.2.8, the set of periodic points is a subset of $C$. ∎

**13.2.10 Corollary** *The set of all the zeroes of all the iterates of $f : [0,1] \to [0,r]$ is a subset of $C$.*

*Proof:* The set of all zeroes of all the iterates of $f : [0,1] \to [0,r]$ is an invariant subset of $[0,1]$ (see Exercise 13.2.12). Hence by Theorem 13.2.8, the set of zeroes is a subset of $C$. ∎

**13.2.11 Corollary** *The set of limits of sequences with elements in $C$ is a subset of $C$.*

*Proof:* The set of limits of sequences with elements in $C$ is an invariant subset of $[0,1]$ (see Exercise 13.2.13). Hence by Theorem 13.2.8, the set of all such limits is a subset of $C$. ∎

Here are a few tips on proving things about sets, for use in the exercises.

**Proofs About Sets**

(a) To prove that $A \subseteq B$ where $A$ and $B$ are sets, show that *if $x \in A$ then $x \in B$.*

(b) To prove something about the union of a family of sets recall that a member of the *union* is in *at least one* of the sets in the family.

On the other hand a member of an *intersection* of sets is in *all* of the sets.

(c) To prove something about $f(A)$ where $A \subseteq \text{dom } f$, recall that

$$f(A) = \{f(x) : x \in\}A.$$

## Exercises 13.2

> In the following exercises,
>
> $f : [0, 1] \to [0, r]$ is a one-hump mapping ($r > 1$),
>
> $C_n$ is the domain of $f^n$,
>
> $C = \cap_{i-1}^{\infty} C_n$.

13.2.1. Which of the following subsets of $[0, 1]$ are invariant sets for a mapping $f : C \to C$?

(a) $\{x_0\}$ where $x_0$ is a fixed point of $f$.

(b) $\{x_0\}$ where $x_0$ is the initial value a prime period two point of $f$.

(c) $\{x_0, x_1, x_2\}$ where $x_0$ is the initial value for an orbit of period 3 for $f$, and $x_1$ and $x_2$ are the next two iterates of $x_0$.

(d) All the points of any orbit $(x_0, x_1, x_2, x_3, \dots)$ of $f$.

13.2.2. Prove each of the following set inclusions, by referring to the appropriate definition (compare with Figure 13.2.1). For all $n$:

(i) $C_n \supseteq C_{n+1}$;

(ii) $C_n \supseteq C$.

13.2.3. Prove that $f(C_{n+1}) \subseteq C_n$, for all $n \in \mathbb{N}$.

[Hint. Use the definition $C_{n+1} = \text{dom } f^{n+1} = \text{dom } f^n \circ f$]

13.2.4. Justify each step in the following proof that the set $C$ is invariant:

$$\begin{aligned} f(C) &= f(\cap_{n=1}^{\infty} C_n) \\ &= f(\cap_{n=1}^{\infty} C_{n+1}) \\ &\subseteq \cap_{n=1}^{\infty} f(C_{n+1}) \\ &\subseteq \cap_{n=1}^{\infty} C_n \\ &= C. \end{aligned}$$

Hence $f(C) \subseteq C$, so that $C$ is an invariant set for $f$.

13.2.5. Show that if $C$ is an invariant set for $f$, then $C$ is also an invariant set for $f^n$ for all $n$.

13.2.6. Find a set $S$ and a mapping $f$ such that $S$ is an invariant set for $f^2$ but not for $f$.

13.2.7. Let $n \in \mathbb{N}$. Prove that if $x \in C_1$ and $f(x) \in C_n$, then $x \in C_{n+1}$.

[Hint. Use the definition $C_{n+1} = \text{dom } f^{n+1} = \text{dom } f^n \circ f$.]

13.2.8. Prove that $f : C \to C$ is onto.

[Hints. Let $y \in C$. Why does $y$ belong to $C_n$ for every $n$?
Why is there an $x \in C_1$ such that $y = f(x)$?
Deduce from Exercise 7 that $x \in C_n$ for all $n$.
Why does this show that $x \in C$?]

13.2.9. Show that if $f : C \to C$ is onto then $f^n$ is onto, for all $n$.

13.2.10. Prove that if $f$ is a one-hump mapping then, for all $m$, each hump
of $f^m$ maps the part of $C$ lying in its base onto the whole of $C$.

[Hints. Let $y \in C$. Why does $y$ belong to $C_n$ for every $n$?
Why is there an $x$ in the base of the hump such that $y = f^m(x)$?
Deduce from Exercise 7 that $x \in C_n$ for all $n$.
Why does this show that
(i) $x \in C$? (ii) $x \in C \cap$ *base of hump*? ]

13.2.11. Show that the set of all periodic points for $f$ is an invariant set for the mapping $f$.

13.2.12. Show that the set of all zeroes of all iterates of $f$ is an invariant set for the mapping $f$.

13.2.13. Show that the set of limits of all sequences of elements from $C$ is an invariant set for the mapping $f$.

## 13.3 WIGGLY ITERATES AND CANTOR SETS

The idea of a mapping $f : [0,1] \to [0,r]$, where $r > 1$, having *wiggly iterates* is suggested by the behaviour, noted in Section 13.1, of the iterates of the tent and the logistic mappings.

**13.3.1 Definition (Wiggly iterates)**

A mapping $f : [0,1] \to [0,r]$ has *wiggly iterates* if:

(i) it is a one hump mapping; and

(ii) the maximum length of the bases of the $2^{n-1}$ humps in $f^n$ approaches 0 as $n$ approaches $\infty$. ∎

Although this definition is formally the same as Definition 7.2.3, the pictures that go with it are different: the bases of the humps now have gaps between them. The following examples illustrate two possibilities.

**13.3.2 Example** *Let $f : [0,1] \to [0,1.0875]$ be the logistic mapping $f = Q_{4.35}$ so that*

$$f(x) = 4.35x(1-x).$$

*The graphs of successive iterates of the one-hump mapping $f$ are shown in Figure 13.3.1.* ∎

**13.3.3 Example** *Let $g : [0,1] \to [0,1.125]$ be the mapping with*

$$g(x) = \frac{4.5x(1-x)}{1+1200x^2(x-0.5)^2(x-1)^2}.$$

*Graphs of successive iterates of the one-hump mapping $g$ are shown in Figure 13.3.1.* ∎

The graphs suggest that, as $n \to \infty$,

*the length of the longest base in $f^n$ will approach* 0, whereas

*the length of the longest base in $g^n$ will approach a number* $\neq 0$.

Thus the graphs suggest that $f$ has wiggly iterates, whereas $g$ does not.

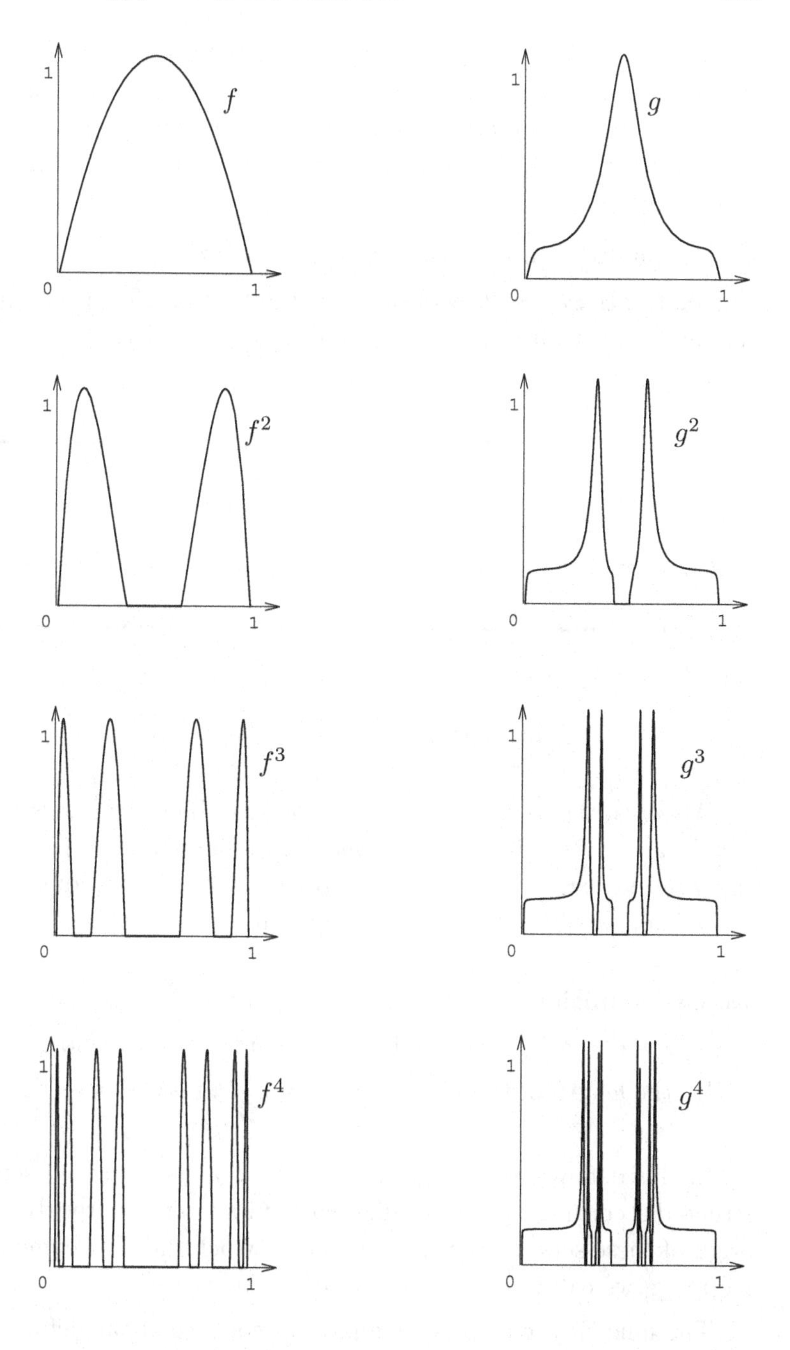


**Fig. 13.3.1** Iterates of $f$ wiggle, but those of $g$ do not.

In Figure 13.3.2 the graphs have been removed, leaving only the bases of the humps. Thus we are left with the domains of successive iterates of each mapping. This diagram reinforces the suggestion that the length of the longest interval in the domain $C_n$ of the $n$th iterate

*approaches* 0 *for the mapping* $f$, *but*

*approaches a number* $\neq 0$ *for the mapping* $g$.

The first of these results will be proved in Section 13.6 as an application of our Test for Chaos; the second in Exercise 13.3.5.

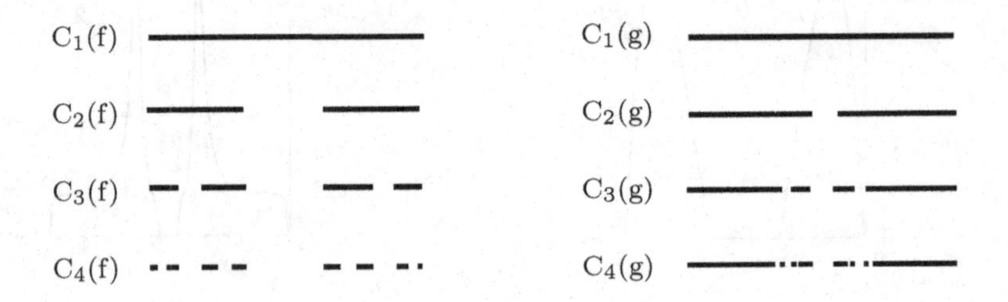


**Fig. 13.3.2** Domains of the iterates of $f$ and of $g$.

We already have an alternate way of describing the *intersection of the domains* of the iterates of a one-hump mapping $C = \bigcap_{n=1}^{\infty} C_n$ as the *largest invariant set* for the mapping. We now introduce a third way of talking about the set $C$ when the mapping has wiggly iterates.

**13.3.4 Definition** A *Cantor set* is any set $\bigcap_{n=1}^{\infty} C_n$ where:

(a) the $C_n$ satisfy the conditions of Lemma 13.2.3; and

(b) the length of the longest interval in $C_n \to 0$ as $n \to \infty$. ■

In this definition, we regard the sets $C_n$ as just any old sets constructed according to certain rules, without reference to the dynamical orign of these sets. In the special case in which the $C_n$'s come from a mapping, we call $C$ *the Cantor set of the mapping.*

The following lemma is a simple consequence of the definitions.

**13.3.5 Lemma** *A one-hump mapping $f : [0,1] \to [0,r]$ has wiggly iterates if and only if the intersection $C$ of the domains of all its iterates is a Cantor set.* ∎

Thus for the above mappings, the largest invariant subset of $f$ is a Cantor set, while that of $g$ is not.

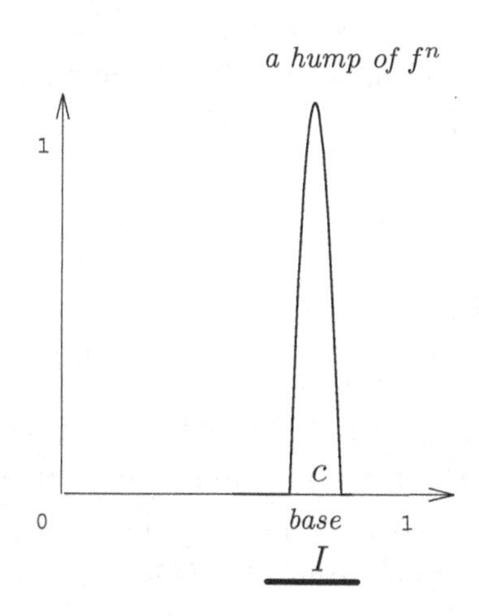


**Fig. 13.3.3** A spike over the open interval $I$ containing $c \in C$.

**13.3.6 Lemma (Spike Lemma)** *Let $f : [0,1] \to [0,r]$ have wiggly iterates ($r > 1$). For every open interval $I \subseteq [0,1]$ which contains a point $c$ of the Cantor set $C$ for $f$, there is a hump of some $f^n$ with base contained in I.*

*Proof:* Let $f$ have wiggly iterates, and let $c \in C$. Let $I$ be an open interval containing $c$.

Since $c \in C$, it follows that $c \in C_n$ for every $n$. Hence $c$ is in a base $I_n$ of a hump of $f^n$ for every $n$.

But the maximum length of the bases $\to 0$ as $n \to \infty$.
Hence the sequences of endpoints approach $c$. Hence the base $I_n \subseteq I$ for all $n$ sufficiently large. ∎

## Exercises 13.3

13.3.1. Let $f : [0, 1] \to [0, r]$, where $r > 1$ be a one-hump mapping, let $C_n$ be the domain of $f^n$, and let $C$ be the intersection of the $C_n$'s.
Match each entry on the left with all appropriate entries on the right.

(a) largest invariant set for $f$ (a′) maximal closed intervals in $C_n$

(b) Cantor set for $f$ (b′) endpoints of bases of humps in $f^n$

(c) bases of humps of $f^n$ (c′) $C$ when $f$ has wiggly iterates

(d) zeroes of $f^n$ (d′) $C_n$

(e) domain of $f^n$ (e′) $\bigcap_{n=1}^{\infty} C_n$

13.3.2. Let $f : [0, 1] \to [0, r]$, where $r > 1$, be a one-hump mapping. In which of the following cases are the pairs of domains $C_2$ and $C_3$ (for $f^2$ and $f^3$ respectively) consistent with Lemma 13.2.3? Give reasons for your answers.

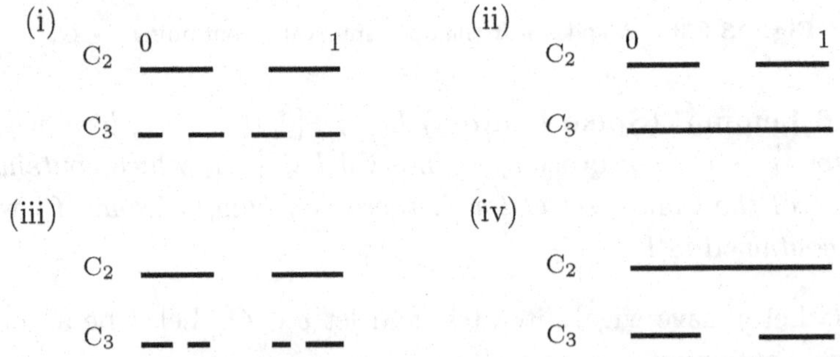

(i) $C_2$ 0 1; $C_3$

(ii) $C_2$ 0 1; $C_3$

(iii) $C_2$; $C_3$

(iv) $C_2$; $C_3$

13.3.3. By using Theorem 13.1.5 and Definition 13.3.4, show that the interesection of the domains of the iterates of the mapping $T = T_6$ is a Cantor set. Describe also the recursive procedure for constructing this particular Cantor set.

[Note. The Cantor set of the tent mapping $T_6$ is often described as *the* Cantor set or the *middle-thirds* Cantor set.]

13.3.4. Let $T : [0,1] \to [0,r]$ be a tent mapping whose graph is an isosceles triangle of height $r > 1$. Show that the parameter $\alpha = 1 - 1/r$ satisfies the inequality $|\alpha| < 1$. From Theorem 13.1.5 (suitably generalized) and Definition 13.3.4, show that the maximal invariant set of $T$ is a Cantor set.

[This set is called the *middle-$\alpha$* Cantor set. Can you explain why?]

13.3.5. Is either of the following 'improved' versions of the Spike Lemma true? Give reasons for your answers.

(a) *Let $f : [0,1] \to [0,r]$ have wiggly iterates ($r > 1$). For every interval $I \subseteq [0,1]$ which contains a point of the Cantor set $C$ for $f$, there is a hump of some $f^n$ with base contained in $I$.*

(b) *Let $f : [0,1] \to [0,r]$ have wiggly iterates ($r > 1$). For every open interval $I \subseteq [0,1]$, there is a hump of some $f^n$ with base contained in $I$.*

13.3.6. Let $g$ be the mapping of Example 13.3.3. The graphs of $g$ and id are shown in Figure 13.3.4.

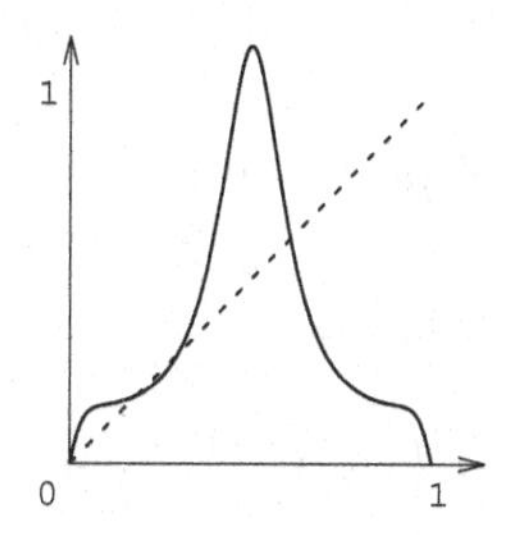


**Fig.13.3.4.** $g$ has an attracting fixed point.

(a) The graphs indicate that $g$ has three fixed points. Which of these is an attractor?

(b) Deduce from Theorem 5.2.1 that $g$ does not have wiggly iterates.

## 13.4 DENSE IN A CANTOR SET

To discuss concepts like *denseness* of periodic points, in a Cantor set $C$, we need to clarify what we mean by the concept of 'distance' between a pair of points in $C$. The appropriate context for this is a metric space.

### Subsets as metric spaces

We already have a formula which defines the distance between two points $x$ and $y$ in $\mathbb{R}$, namely

$$d(x, y) = |x - y|.$$

The mapping $d$ is called the *usual metric* for $\mathbb{R}$. Since $C \subseteq \mathbb{R}$ we can define the distance between two points $x$ and $y$ in $C$ by using the same formula. Thus we are using the distance between points in $\mathbb{R}$ to obtain the distance between points in $C$. We can express this by saying that we have obtained a metric for $C$ by restriction of the metric $d$ for $\mathbb{R}$.

If $r > 0$ the open interval $(a - r, a + r)$ in $\mathbb{R}$ can be written as

$$I = \{x \in \mathbb{R} : |x - a| < r\},$$

which is just the set of points in the metric space $\mathbb{R}$, with the usual metric, whose distance from $a$ is less than $r$. In the metric space $C$, however, the analogous set is

$$\{x \in C : |x - a| < r\}$$

which is equal to $I \cap C$, where $I$ is the above open interval in $\mathbb{R}$.

In the metric space $C$, sets of the form $I \cap C$ play a role which is analogous to that played by the open intervals in $\mathbb{R}$. Inasmuch as open intervals in $\mathbb{R}$ are non-empty sets, we complete the analogy by imposing the condition that $I \cap C \neq \varnothing$.

The above analogy motivates the following definition of *denseness* in the metric space $C$.

**13.4.1 Definition** A set $D \subseteq C$ is said to be *dense in $C$* if:

*for every interval $I$ such that $I \cap C \neq \varnothing$,*

*there is a point of $D$ which is in $I \cap C$.* ∎

The above definition is all that you need in order to be able to proceed with Section 13.5. The following lemmas illustrate how the definition is used. They provide a result for the Cantor set $C$ which is analogous to Lemma 7.3.2 for the interval $[0, 1]$. This result will be used in Section 13.6.

**13.4.2 Lemma (Wiggly implies dense zeroes)**
*Let $f : [0, 1] \to [0, r]$, where $r > 1$. If the mapping $f$ has wiggly iterates, then the set $S$ of all zeroes of all iterates of $f$ is dense in $C$, the Cantor set of $f$.*

*Proof:* This is a simple exercise on the use of the Spike Lemma. See Exercise 13.4.8 for details. ∎

To prove the converse of Lemma 13.4.2 we need the following definition and theorem from Analysis.

| **Cluster Points** |
|---|
| Let $\{s_n\}_{n=1}^{\infty}$ be a sequence of points in the interval $[0, 1]$. A point $\ell \in \mathbb{R}$ is a *cluster point* of the sequence if every open interval containing $\ell$ also contains $s_n$ for infinitely many $n$. |

| **Sequential Compactness** |
|---|
| The interval $[0, 1]$ has the property of *sequential compactness*: which means that every sequence of points in $[0, 1]$ has a cluster point in $[0, 1]$. |

**13.4.3 Lemma (Dense zeroes implies wiggly)**
*Let $f : [0,1] \to [0,r]$, where $r > 1$. If the set $S$ of all zeroes of all iterates of $f$ is dense in the set $C = \bigcap_{n=1}^{\infty} C_n$ (the intersection of the domains of the iterates of $f$) then the mapping $f$ has wiggly iterates*

.

*Proof:* We prove the contrapositive form of the lemma.

Suppose the mapping does not have wiggly iterates.

We aim to show that the zeroes are not dense in $C$;

that is, we aim to find an open interval $I$
such that $I \cap C$ contains no zeroes.

As the iterates are not wiggly, there is a sequence of intervals $\{J_n\}_{n=1}^{\infty}$ such that

(a) every $J_n$ is the base of a hump of $f^n$ and
(b) the length of $J_n$ does not approach 0 as $n \to \infty$.

Hence there is a $p > 0$ such that every interval $J_n$ has length $\geq p$.

The mid points of the intervals $J_n$ form a sequence in $[0,1]$.

The sequence of mid-points has a cluster point $\ell \in [0,1]$ (since $[0,1]$ is sequentially compact).

Hence the mid-point of $J_n$ is in the interval $I = (\ell - p/4, \ell + p/4)$, for some $n$ which can be chosen arbitrarily large.

Hence $I \subseteq J_n$ for any such $n$.

But $J_n$, without its end-points, contains no zeroes of $f^n$ (being the base of a hump of $f^n$).

Thus $I$ contains no zeroes of $f^n$, for some $n$ arbitrarily large.

But since zeroes are preserved under iteration, this holds for all $n$.

Also $I \subseteq J_n \subseteq C_n$ for some $n$ arbitrarily large.

Hence $I \subseteq C_n$ for all $n$ (since $C_1 \supseteq C_2 \supseteq \cdots \supseteq C_{n-1} \supseteq C_n$).

Hence $I \subseteq C$.

Thus there is an interval $I$ such that $I \cap C \neq \varnothing$ and $I \cap C$ contains no zeroes of any iterate of $f$. Thus the zeroes are not dense in $C$. ▪

## Exercises 13.4

13.4.1. Prove from the definition that $C$ is dense in $C$ for every set $C \subseteq \mathbb{R}$.

13.4.2. Write down the negation of the statement which defines the density of a set $D \subseteq C$ in $C \subseteq \mathbb{R}$.

> In the remaining exercises the mappings are all one-hump mappings from $[0, 1]$ into $[0, r]$ where $r > 1$.

13.4.3. Let $C$ be the Cantor set of a one-hump mapping with wiggly iterates. Give an example of a sequence whose elements are dense in $C$.

13.4.4. Let $C(f)$ and $C(g)$ be the maximal invariant sets for the one-hump mappings $f$ and $g$, respectively. Show that $C(f) \cap C(g)$ contains at least two elements of $[0, 1]$.

13.4.5. Let $C$ be the maximal invariant set for a one-hump mapping $f$. Give an example of a set $D \subseteq C$ which is not dense in $C$.

13.4.6. Let $C$ be the maximal invariant set of a one-hump mapping. Show that $C$ is not dense in $[0, 1]$.

13.4.7. Let $C$ be the Cantor set of a mapping $f$ with wiggly iterates. Show that $C$ does not include any intervals.

13.4.8. Prove Lemma 13.4.2.

[Hints. Assume that $f$ has wiggly iterates and let $I \subseteq [0, 1]$ be an open interval with $I \cap C_n \neq \varnothing$. Then:

Why does $I$ contain a base of a hump of some $f^n$?

How does this show that the zeroes are dense in $C$ ? ]

13.4.9. Give an example of a sequence of points in $[0, 1]$ in each case:

(i) The sequence has just one cluster point in $[0, 1]$.

(ii) The sequence has just two cluster points in $[0, 1]$.

(ii) The sequence has exactly $k$ cluster points in $[0, 1]$.

## 13.5 CHAOS ON CANTOR SETS

Chaos, in the sense of Devaney, means sensitive dependence on initial conitions, transitivity and a dense set of periodic points. We show how to adapt these notions to mappings from a Cantor set $C$ into itself. We then show that if $f : [0,1] \to [0,r]$ has wiggly iterates, then its restriction to its Cantor set $f : C \to C$ is a chaotic mapping.

### Sensitive dependence

**13.5.1 Definition** A mapping $f : C \to C$ has *sensitive dependence at $x \in C$* if the following condition holds for some $\delta > 0$: for each open interval $I$ containing $x$, there is a $y \in I \cap C$ and an $n \in \mathbb{N}$ such that

$$|f^n(x) - f^n(y)| \geq \delta.$$

The number $\delta$ is called a sensitivity constant for the mapping. ■

**13.5.2 Theorem (Wiggly implies sensitive dependence)** *If $f : [0,1] \to [0,r]$ has wiggly iterates and $C$ is its Cantor set, then the restricted mapping $f : C \to C$ has sensitive dependence everywhere (with sensitivity constant $\frac{1}{2}$ if the mapping is symmetric).*

*Sketch of Proof:* The proof is similar to that of Theorem 9.1.1. The additional ideas needed are

(i) the endpoints of the base of a hump are in the Cantor set $C$ (by Corollary 13.2.10),

(ii) the point where a hump assumes the value 1 is also in the Cantor set since if $f^n(x) = 1$ then $f^{n+1}(x) = f(1) = 0$. ■

### Transitivity

**13.5.3 Definition** A mapping $f : C \to C$ is *transitive* if, for every pair of open intervals $I$ and $J$ having non-empty intersection with $C$, there is an $n \in \mathbb{N}$ such that

$$f^n(I \cap C) \cap (J \cap C) \neq \varnothing.$$ ■

**13.5.4 Theorem (Wiggly implies transitive)**
*If $f : [0, 1] \to [0, r]$ has wiggly iterates then $f : C \to C$ is transitive*

*Sketch of Proof:* The proof is similar to that of Lemma 9.2.2. The additional fact needed is: the base of a hump of $f^n$ contains a copy of the Cantor set of the mapping, and this copy is mapped onto $C$ by this hump of $f^n$ (by Theorem 13.2.7).

## Dense periodic points

**13.5.5 Theorem (Wiggly implies dense periodic points)**
*If $f : [0, 1] \to [0, r]$ has wiggly iterates then the set of periodic pionts of $f : C \to C$ is dense in $C$.*

*Sketch of proof:* The proof follows that of Lemma 9.3.1. The additional fact needed is that all the periodic points of $f : [0, 1] \to [0, r]$ are in the Cantor set $C$ (by Corollary 13.2.9). ■

Thus, to prove that a one-hump mapping $f : [0, 1] \to [0, r]$ has chaotic behaviour on its Cantor set, it is sufficient to show that it has wiggly iterates.

**13.5.6 Example** *Show that the tent-mapping $T_6$ has chaotic behaviour on its Cantor set.*

*Solution:* $T_6$ is a one-hump mapping. By Theorem 13.1.5, this mapping has wiggly iterates. ■

Usually, it is not easy to prove that a mapping has wiggly iterates. In the next section we give test which can be applied to a large class of mappings.

## Exercises 13.5

13.5.1. Complete the proofs of the three theorems stated in the text.

13.5.2. Adapt the exercises set for Sections 9.1, 9.2, and 9.3.

## 13.6 TESTS FOR CHAOS AND CONJUGACY

The theorems proved in Sections 10.3 and 12.3, concerned the chaotic behaviour of mappings from the interval $[0,1]$ into itself. Straightforward modifications of these theorems give results which are valid for mappings which send $[0,1]$ onto $[0,r]$, where $r > 1$. The changes which need to be made to the proofs are also straightforward. Hence the proofs are left as exercises.

**13.6.1 Theorem (Test for wiggly iterates)**
*Let $r > 1$ and let $f : [0,1] \to [0,r]$ be a symmetric one-hump mapping. If $f'(0) > 1$ and $f$ has no woggles, then $f$ has wiggly iterates.*

*Proof:* This is similar to that of Theorem 10.3.1. ∎

**13.6.2 Theorem (Test for chaos)**
*Let $f : [0,1] \to [0,r]$ be a symmetric one-hump mapping. If $f'(0) > 1$ and if $f$ has negative Schwarzian derivative (except possibly at $\frac{1}{2}$), then the restriction of $f$ to its Cantor set has chaotic behaviour.*

*Proof:* Similar to that of Theorem 10.3.2. ∎

**13.6.3 Example** *Show that the logistic mapping $Q_\mu$ is chaotic for $\mu > 4$.*

*Solution:* $Q_\mu$ maps $[0,1]$ onto $[0,\mu/4]$ and $\mu/4 > 1$ when $\mu > 4$. Also $Q_\mu$ is a symmetric one-hump mapping with negative Schwarzian derivative. By Theorem 13.6.2, $Q_\mu$ is chaotic on its Cantor set. ∎

**13.6.4 Theorem (Conjugacy)**
*Let $f : [0,1] \to [0,r]$ and $g : [0,1] \to [0,s]$ where $r > 1$ and $s > 1$. Suppose that both $f$ and $g$ have wiggly iterates, negative Schwarzian derivatives and that their Cantor sets are $C(f)$ and $C(g)$ respectively. There is then a homeomorphism $h : C(f) \to C(g)$ such that*

$$h \circ f = g \circ h.$$

*Proof:* This can be obtained by straightforward modifications of the material in Section 12.3. ■

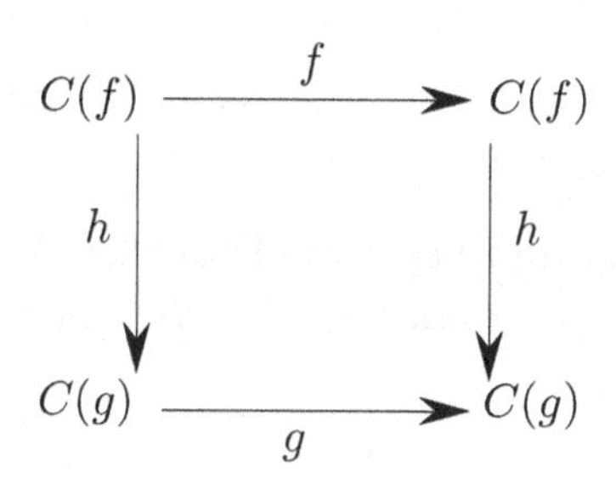


**Fig.16.3.1.** Commutative diagram for conjugacy.

**13.6.5 Corollary** *Let $f : [0,1] \to [0,r]$, where $r > 1$. If is a symmetric one-hump mapping with negative Schwarzian derivatve, then the restriction of $f$ to its Cantor set is conjugate to the restriction of the tent mapping $T_6$ to its Cantor set.* ■

---

## Additional reading for Chapter 13

Section 2.1 of [PJS] discusses Cantor sets and their history at some length and goes on to explain how the Middle Thirds Cantor Set arises as the invariant set for the Tent map. The authors use the term 'prisoner set' instead of invariant set. Other accounts of invariant Cantor sets for tent mappings are given in Section 9.3 of [De2], Section 6.3 of [PMJPSY] and Section 6.8 of [AP].

Accounts of invariant Cantor sets for mappings in the logistic family are given in Section 1.5 of [De1] Section 2.4 of [Gul] and Chapter 8 of [Ho].

Many of the above texts go on to discuss various two-dimensional fractals. The first chapter of [Ed] discusses a number of interesting approaches to the construction of Cantor sets. Later chapters discuss two-dimensional fractals.

One example of a chaotic two-dimensional invariant set closely related to the Cantor set is called "Smale's Horseshoe". This is discussed in Section 6.8 of [AP], Section 2.3 of [De1] and in [BD].

## References


[AP] D.K. Arrowsmith and C.M. Place, *Dynamical Systems: Differential Equations, Maps and Chaotic Behaviour,* Chapman and Hall, 1992.

[BD] John Banks and Valentina Dragan, "Smale's Horseshoe Map Via Ternary Numbers", *Siam Review,* **36** (June 1994), 265-271.

[De1] Robert L. Devaney, *An Introduction to Chaotic Dynamical Systems: Second Edition,* Addison-Wesley, Menlo Park California, 1989.

[De2] Robert L. Devaney, *Chaos, Fractals and Dynamics, Computer Experiments in Mathematics,* Addison-Wesley, New York, 1990.

[Ed] Gerald Edgar, *Measure, Topology and Fractal Geometry,* Springer-Verlag, New York, 1990.

[Gul] Denny Gulick, *Encounters with Chaos,* McGraw-Hill, Inc., 1992.

[Ho] Richard Holmgren, *A First Course in Discrete Dynamical Systems* Springer-Verlag, 1994.

[PJS] Hans-Otto Peitgen, Hartmut Jürgens and Dietmar Saupe, *Chaos and Fractals, New Frontiers of Science*, Springer-Verlag, New York, 1992.

[PMJPSY] Hans-Otto Peitgen, Evan Maletsky, Hartmut Jürgens, Terry Perciante, Dietmar Saupe, and Lee Yunker, *Fractals for the Classroom: Strategic Activities Volume Two,* Springer-Verlag, 1992. (two volumes)

# Index

*Definitions and statements of theorems and lemmas are shown in* **bold.** *An f denotes a reference in a footnote.*

Printed in the United States
By Bookmasters